THE STORY OF THE SUN, MOON AND STARS

THE STORY OF THE SUN, MOON AND STARS

AGNES GIBERNE

GYAN

Published by

Gyan Publishing House
5, Ansari Road
Daryaganj, New Delhi-110002
Phone: 011-47034999, 9811692060
E-mail: books@gyanbooks.com

Distribution Network
gyanbooks.com
India, USA, Canada, UK, Australia

ISBN: 978-93-7316-219-5 (PB)

First Published, 1898

2nd Impression 2025

Printed at: Gyan Press, Delhi.

THE STORY OF THE SUN, MOON AND STARS
Author: AGNES GIBERNE

HIPPARCHUS IN THE OBSERVATORY OF ALEXANDRIA.

THE STORY

OF THE

SUN, MOON, AND STARS

BY

AGNES GIBERNE

AUTHOR OF "THE STARRY SKIES," "THE WORLD'S FOUNDATIONS," "RADIANT SUNS," ETC.

WITH COPIOUS ADDITIONS

Fully Illustrated

CINCINNATI
NATIONAL BOOK COMPANY

INTRODUCTION.

By Charles Pritchard, M.A., F.R.S.,

Professor of Astronomy in the University of Oxford.

The pages of this volume on a great subject were submitted to my criticism while passing through the press, with a request from a friend that I would make any suggestions which might occur to me for its improvement. Naturally, such a request was entertained, in the first instance, with hesitation and misgiving. But after a rapid perusal of the first sheet, I found my interest awakened, and then gradually secured; for the book seemed to me to possess certain features of no ordinary character, and, in my judgment, held out the promise of supplying an undoubted want, thus enabling me to answer a question which I have been often asked, and which had as often puzzled me, to the effect, "Can you tell me of any book on astronomy suited to beginners?" I think just such a book is here presented to the reader; for the tale of the stellar universe is therein told with great simplicity, and perhaps with sufficient completeness, in an earnest and pleasant style, equally free, I think, from inaccuracy or unpardonable exaggeration. We have here the outlines of elementary astronomy, not merely detailed without mathematics, but to a very great extent expressed in untechnical language. Success in such an attempt I regard as a considerable feat, and one of much practical utility.

For the science of astronomy is essentially a science of great magnitude and great difficulty. From the time of Hipparchus, some century and a half before the Christian era, down to the present day, the cultivation of astronomy has severely taxed the minds of a succession of men endowed with the rarest genius. The facts and the truths of the science thus secured have been of very slow accretion; but like all other truths, when once secured and thoroughly understood, they are found to admit of very simple verbal expression, and to lie well within the general comprehension, and, I may safely add, within the sympathies of all educated men and women.

Thus the great astronomers, the original discoverers of the last twenty centuries, have labored, each in his separate field of the vast universe of nature, and other men, endowed with other gifts, have entered on their labors, and by systematizing, correlating, and simplifying the expression of their results, have brought the whole within the grasp of cultivated men engaged in other branches of the varied pursuits of our complicated life. It is in this sort of order that the amelioration and civilization of mankind have proceeded, and at the present moment are, I hope and believe, rapidly proceeding.

It was, I suspect, under this point of view, though half unconsciously so, that my attention was arrested by the book now presented to the reader; for we have here many of the chief results of the laborious researches of such men as Ptolemy, Kepler, Newton, Herschel, Fraunhofer, Janssen, Lockyer, Schiaparelli, and others—no matter where accumulated or by whom recorded—filtered through the mind of a thoughtful and cultured lady, and here presented to other minds in the very forms wherein they have been assimilated and pictured in her own. And these forms and pictures are true. It is in this way that the intellectual "protoplasm" of the human mind is fostered and practically disseminated.

And, then, there is still another point of view from which this general dissemination of great truths in a simple style assumes an aspect of practical importance. I allude to the influences of this process on the imaginative or poetic side of our complex nature.

Wordsworth, in one of his prefaces, has stated so clearly the truth on this subject that I can not do better than give his words. "If the time should ever come," he says, "when what is now science becomes familiarized to men, then the remotest discoveries of the chemist, the botanist, the mineralogist, will be as proper objects of the poet's art as any upon which it can be employed. He will be ready to follow the steps of the man of science; he will be at his side, carrying sensation into the midst of the objects of science itself. The poet will lend his divine spirit to aid the transfiguration, and will welcome the being thus produced as a dear and genuine inmate of the household of man."

It is for reasons such as here stated that I heartily commend this book to the attention of those who take an interest in the advancement of the intellectual progress and culture of society. The story of the Kosmos is told by the authoress in her own language and after her own method. I believe, as I have said, the story is correct.

OXFORD UNIVERSITY.

CONTENTS.

LIST OF ILLUSTRATIONS.

THE STORY

OF THE

SUN, MOON, AND STARS.

CHAPTER I.

THE EARTH ONE OF A FAMILY.

WHAT is this earth of ours?

Something very great—and yet something very little. Something very great, compared with the things upon the earth; something very little, compared with the things outside the earth.

And as our journeyings together are for a while to be away from earth, we shall find ourselves obliged to count her as something quite small in the great universe, where so many larger and mightier things are to be found,—if indeed they are mightier. Not that we have to say good-bye altogether to our old home. We must linger about her for a while before starting, and afterwards it will be often needful to come back, with speed swifter than the flight of light, that we may compare notes on the sizes and conditions of other places visited by us.

But first of all: what *is* this earth of ours?

It was rather a pleasant notion which men held in olden days, that we—that great and important "We" which loves to perch itself upon a height—stood firm and fixed at the very center of everything. The earth was supposed to be a vast flat plain, reaching nobody could tell how far. The sun rose and set for us alone; and the thousands of stars twinkled in the sky at night for nobody's good except ours; and the blue sky overhead was a crystal covering for the men of our earth, and nothing more. In fact, people seem to have counted themselves not merely to have had a kind of kingship over the lower animals of our earth, but to have been kings over the whole universe. Sun, stars, and sky, as well as earth, were made for man, and for man only.

This was the common belief, though even in those olden days there were *some* who knew better. But the world in general knows better now.

Earth the center of the universe! Why, she is not that of even the particular family in the heavens to which she belongs. For we do not stand alone. The earth is one of a family of worlds, and that family is called THE SOLAR SYSTEM. And so far is our earth from being the head of the family, that she is not even one of the more important members. She is merely one of the little sisters, as it were.

Men not only believed the earth, in past days, to be at the center of the universe; but also they believed her to remain there without change. Sun, moon, stars, planets, sky, might move; but never the earth. The solid ground beneath their feet, *that* at least was firm. Every day the sun rose and set, and every night the stars, in like manner, rose and set. But this was easily explained. We on our great earth stood firm and still, while sun, moon, stars, went circling round us once in every twenty-four hours, just for our sole

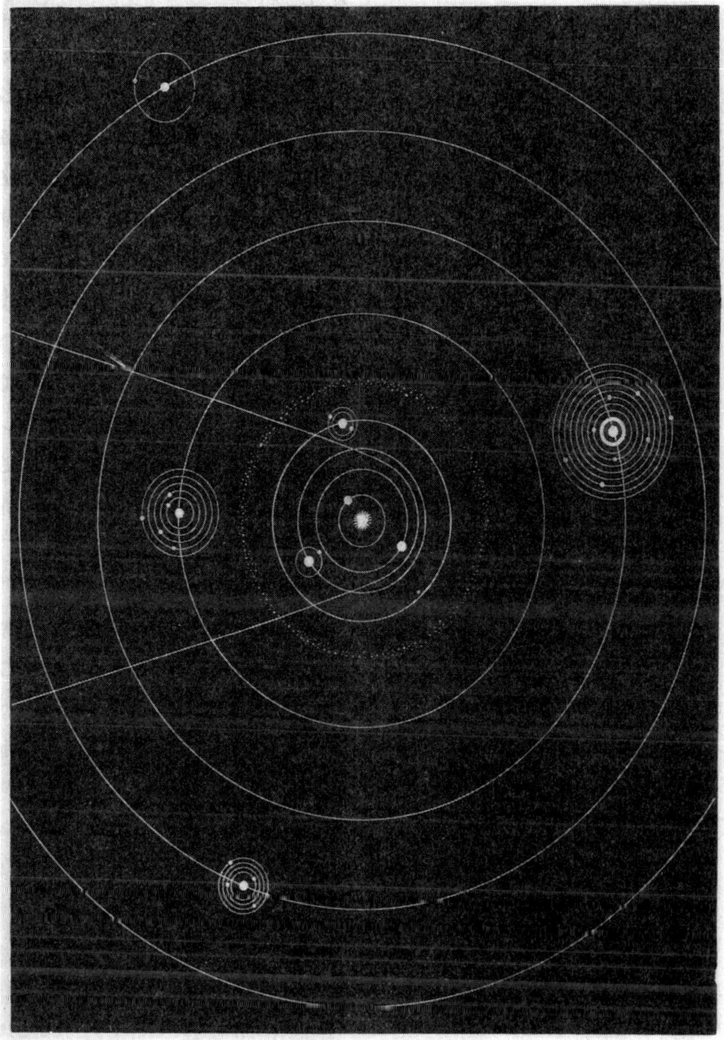

THE SOLAR SYSTEM.

and particular convenience. What an important personage man must have felt himself then in God's great universe! Once again, we know better now!

For it is the earth that moves, and not the sun; it is the earth that moves, and not the stars. The daily movements of sun and stars, rising in the east, traveling over the sky, and setting in the west, are no more real movements on their part than, when we travel in a railway-train, the seeming rush of hedges, telegraph posts, houses, and fields is real. They are fixed and we are moving; yet the movement appears to us to be not ours but theirs.

Still more strongly would this appear to be the case if there were no noise, no shaking, no jarring and trembling, to make us feel that we are not at rest. Sometimes when a train begins to move gently out of a station, from among other trains, it is at the first moment quite impossible to say whether the movement belongs to the train in which we are seated or to a neighboring train. And in the motions of our earth there is no noise, no shaking, no jarring—all are rapid, silent, and even.

If you were rising through the air in a balloon, you would at first only know your own movement by seeing the earth seem to drop away from beneath you. And just so we can only know the earth's movements by seeing how worlds around us *seem* to move in consequence.

When I speak of the Universe, I mean the whole of God's mighty creation as far as the stars reach. Sometimes the word is used in this sense, and sometimes it is used only for a particular part of creation nearer to us than other parts. At present, however, we will put aside all thought of the second narrower meaning.

The wisest astronomer living can not tell us how

far the stars reach. We know now that there is no firm crystal covering over our heads, dotted with bright points here and there; but only the wide open sky or heaven, containing millions of stars, some nearer, some farther, some bright enough to be seen by us all, some only visible through a telescope.

People talk often of the stars being "set in space;" and the meaning of "space" is simply "room." Where you are must be space, or you could not be there. But it is when we get away from earth, and travel in thought through the wide fields of space, where God has placed his stars, that we begin to feel how vast it is, and what specks we are ourselves—nay, what a speck our very earth is, in this great and boundless creation.

For there is no getting to the borders of space. As one telescope after another is made, each one stronger in power and able to reach farther than the last, still more and more stars are seen, and yet more and more behind and beyond, in countless millions.

It is the same all round the earth. The old notion about our world being a flat plain has been long since given up. We know her now to be a round globe, not fixed, but floating like the stars in space.

When you look *up* into the sky, you are looking exactly in the opposite direction from where you would be looking *up* if you were in Australia. For Australia's "up" is our "down," and our "down" is Australia's "up." Or, to put it more truly, "up" is always in the direction straight away from this earth, on whatever part of it you may be standing; and "down" is always towards the center of the earth.

2

All round the globe, in north, south, east, west, whether you are in Europe, Asia, Africa, or America, though you will see different stars in certain different quarters of the earth, still overhead you will find shining countless points of light.

And now, what are these stars? This is a matter on which people are often confused, and on which it is well to be quite clear beforegoing one step farther.

Some of the stars you have most likely often noticed. The seven chief stars of the Great Bear are known to a large number of people; and there are few who have not admired the splendid constellation ot Orion. Perhaps you also know the W-shape of Cassiopeia, and the brilliant shining of Sirius, and the soft glimmer of the Pleiades.

The different constellations or groups of principal stars have been watched by men for long ages past. They are called the fixed stars, for they do not change. How many thousands of years ago they were first arranged by men into these groups, and who first gave them their names, we can not tell.

True, night by night, through century after century, they rise, and cross the sky, and set. But those are only seeming movements. Precisely as the turning round of the earth upon her axis, once in every twenty-four hours, makes the sun appear to rise and cross the sky and set, in the day-time; so also the same turning of the earth makes the stars appear to do the same in the night-time.

There is another seeming movement among the stars, which is only in seeming. Some come into view in summer which can not be seen in winter; and some come into view in winter which can not be seen in summer. For the sun, moving on his pathway through the sky, hides those stars which shine with him in the day-time. The zodiac is an imaginary belt in the heavens, sixteen or eighteen degrees wide, containing the twelve constellations through which the sun passes. And as he passes from point to point of his pathway, he constantly conceals from us fresh groups of stars by day, and allows fresh groups to appear by night. Speaking generally, however, the stars remain the same year after year, century after century. The groups may still be seen as of old, fixed and unchanging.

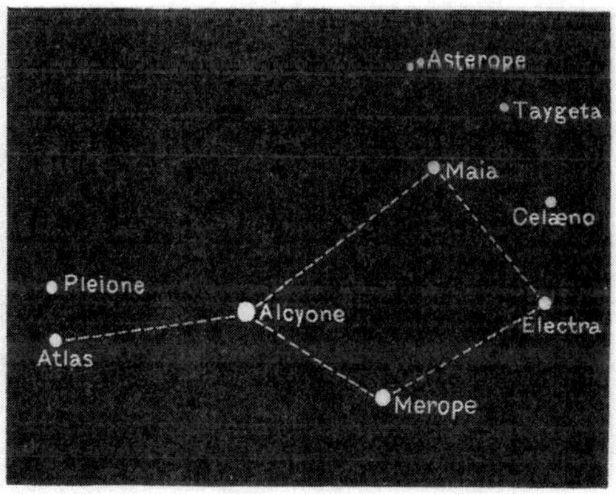

THE PLEIADES.

What are these stars? Stars and planets have both been spoken about. There is a great difference between the two.

Perhaps if you were asked whether the sun is most like to a star or a planet, you would be rather at a loss; and many who have admired the brilliant evening star, Venus, often to be seen after sunset, would be surprised to learn that the evening star is in reality no star at all.

A star is a sun. Our sun is nothing more nor less than a star. Each one of the so-called "fixed stars," that you see shining at night in the sky, is a sun like our sun; only some of the stars are larger suns and some are smaller suns than ours.

The main reason why our sun looks so much larger and brighter than the stars is, that he is so very much nearer to us. The stars are one and all at enormous distances from the earth. By and by we will go more closely into the matter of their distance, compared with the distance of the sun.

At present it is enough to say that if many of the stars were placed just as near to us as our sun is placed, they would look just as large and bright; while there are some that would look a great deal larger and brighter. And if our sun were to travel away from us, to the distance of the very nearest of the little twinkling stars, he would dwindle down and down in size and brilliancy, till at last we should not be able to tell him apart from the rest of the stars.

I have told you that the stars are called "fixed" because they keep their places, and do not change from age to age. Though the movement of the earth makes them seem all to sweep past every night in company, yet they do not travel in and out among one another, or backwards and forwards, or from side to side. At all events, if there be such changes, they are so slow and so small as to be exceedingly difficult to find out. Each group of stars keeps its own old shape, as for hundreds of years back.

But among these fixed stars there are certain stars which do go to and fro, and backwards and forwards. Now they are to be seen in the middle of one constellation, and now in the middle of another. These restless stars were long a great perplexity. Men named them Planets or Wanderers.

We know now that the planets are in reality not stars at all, and also that they are not nearly so far away from us as the fixed stars. In fact, they are simply members of our own family—the Solar System. They are worlds, more or less like the world we live in; and they travel round and round the sun as we do, each more or less near to him; and they depend upon him for heat and light, in more or less the same manner as ourselves.

Therefore, just as our sun is a star, and stars are suns, so our earth or world is a planet, and planets are worlds. EARTH is the name we give to that particular world or planet on which we live. Planets may generally be known from stars by the fact that they do not twinkle. But the great difference between the two lies in the fact that a star shines by its own radiant, burning light, whereas a planet shines merely by light reflected or borrowed from the sun.

But how were the earth and the other worlds made? Let us imagine an immense gaseous mass placed in space. Attraction is a force inherent in every atom of matter. The denser portion of this mass will insensibly attract towards it the other parts, and in the slow fall of the more distant molecules towards this more attractive region, a general motion is produced, incompletely directed towards this center, and soon involving the whole mass in the same motion of rotation. The simplest form of all, even in virtue of this law of attraction, is the spherical form. It is that which a drop of water takes, and a drop of mercury if left to itself.

The laws of mechanics show that, as this gaseous mass condenses and shrinks, the motion of rotation of the nebula is accelerated. In turning, it becomes flattened at the poles, and gradually takes the form of an immense lens-shaped mass of gas. It has begun to turn so quickly as to develop at its exterior circumference a centrifugal force superior to the general attraction of the mass, as when we whirl a sling. The inevitable consequence of this excess is a rupture of the equilibrium, which detaches an external ring. This gaseous ring will continue to rotate in the same time and with the same velocity; but the nebulous mother will be henceforth detached, and will continue to undergo progressive condensation and acceleration of motion. The same feat will be reproduced as often as the velocity of rotation surpasses that by which the centrifugal force remains inferior to the attraction. It may have happened also that secondary centers of condensation would be formed even in the interior of the nebula.

In our system the rings of Saturn still subsist.

The successive formation of the planets, their situation near the plane of the solar equator, and their motions of translation round the same center, are explained by the theory which we are discussing. The most distant known planet, Neptune, would be detached from the nebula at the epoch when this nebula extended as far as the planet, out to nearly three thousand millions of miles, and would turn in a slow revolution requiring a period of 165 years for its accomplishment. The original ring could not remain in the state of a ring unless it was perfectly homogeneous and regular; but such a condition is, so to say, unrealizable, and it did not delay in condensing itself into a sphere. Successively, Uranus, Saturn, Jupiter, the army of small planets, Mars, would thus be detached or formed in the interior of this same nebula. Afterwards came the earth, of which the birth goes back to the epoch when the sun had arrived at the earth's present position. Venus and Mercury would be born later. Will the sun give birth to a new world? This is not probable.

CHAPTER II.

THE HEAD OF OUR FAMILY.

PEOPLE began very early in the history of the world to pay close attention to the sun. And no wonder. We owe so much to his heat and light that the marvel would be if men had not thought much about him.

Was the sun really any larger than he looked, and if so, how much larger was he? And what was his distance from the earth? These were two of the questions which puzzled our ancestors the longest. If once they could have settled exactly how far off the sun really was, they could easily have calculated his exact size; but this was just what they could not do.

So one man supposed that the sun must be quite near, and very little larger than he looked. Another thought he might be seventy-five miles in diameter. A third ventured to believe that he was larger than the country of Greece. A fourth was so bold as to imagine that he might even outweigh the earth herself.

After a while many attempts were made to measure the distance of the sun; and a great many different answers to this difficult question were given by different men, most of them very wide of the mark. It is only of late years that the matter has been clearly settled. And, indeed, it was found quite lately that a mistake of no less than three millions of miles had been made, notwithstanding all the care and all the attention given. But though three millions of miles sounds a great deal, yet it is really very little—only a tiny portion of the whole.

For the distance of the sun from the earth is no less than about NINETY-THREE MILLIONS OF MILES. Ninety-three millions of miles! Can you picture that to yourself? Try to think what is meant by a thousand miles. Our earth is eight thousand miles in diameter. In other words, if you were to thrust a gigantic knitting-needle through her body, from the North Pole to the South Pole, it would have to be about eight thousand miles long.

To reach the thought of one million, you must picture *one thousand times one thousand.* Our earth is about twenty-five thousand miles round. It you were to start from the mouth of the River Amazon, in South America, and journey straight round the whole earth on the equator, till you came back to the same point, you would have traveled about twenty-five thousand miles. But that would be a long way off from a million miles. You would have gone only once round the earth. Now a cord one million miles in length could be wrapped, not once only, but *forty times*, round and round the earth. And when you have managed to reach up to the thought of one million miles, you have then to remember that the sun's distance is ninety-three times as much again. So, to picture clearly to ourselves the actual meaning of "ninety-three millions of miles" is not so easy.

Suppose it were possible to lay a railroad from here to the sun. If you could journey thither in a perfectly straight line, at the rate of thirty miles an hour, never pausing for one single minute, night or day, you would reach the sun in about three hundred and forty-six years.

Thirty miles an hour is a slow train. Suppose we double the speed, and make it an express train, rushing along at the pace of sixty miles an hour. Then you might hope to reach the end of your journey in one hundred and seventy-three years. If you had quitted this earth early in the eighteenth century, never stopping on your way, you would be, just about now, near the end of the nineteenth, arriving at the sun.

So much for the sun's distance from us. Now as to his size.

I have already mentioned that our earth's diameter —that is, her *through measure*, as, for instance, the line drawn straight from England through her center to New Zealand—is about eight thousand miles. This sounds a good deal. But what do you think of the diameter of the sun being no less than *eight hundred and fifty-eight thousand miles?* The one is eight thousand miles, the other over eight hundred thousand!

Suppose you had a long slender pole which would pass through the middle of the earth, one end just showing at the North Pole and the other at the South Pole. You would need more than a hundred and eight of such poles, all joined together, to show the diameter of the sun.

The sun seems not to be made of nearly such heavy materials as the earth. He is what astronomers call less "dense," less close and compact in his make, just as wood is less dense and heavy than iron. Still, his size is so enormous, that if you could have a pair of gigantic scales, and put the sun into one scale and the earth with every one of her brother and sister planets into the other, the sun's side would go down like lightning. He would be found to weigh seven hundred and fifty times as much as all the rest put together. And, it would take more than twelve hundred thousand little earths like ours, rolled into one huge ball, to make a globe as large as the sun.

In the beginning of the seventeenth century a man named Fabricius was startled by the sight of a certain black spot upon the face of the sun. He watched till too dazzled to look any longer, supposing it to be a small cloud, yet anxious to learn more. Next day the spot was there still, but it seemed to have moved on a little way. Morning after morning this movement was found to continue, and soon a second spot, and then a third spot, were observed creeping in like manner across the sun. After a while they vanished, one at a time, round his edge, as it were; but after some days of patient waiting on the part of the lookers-on, they appeared again at the opposite edge, and once more began their journey across.

Fabricius seems to have been the first, but he was not the last, to watch sun-spots. Many astronomers have given close attention to them. Modern telescopes, and the modern plan of looking at the sun through darkened glass, have made this possible in a way that was not possible two or three hundred years ago.

The first important discovery made through the spots on the sun, was that the sun turns round upon his axis, just in the same manner that the earth turns round upon hers. Instead of doing so once in the course of each twenty-four hours, like the earth, he turns once in the course of about twenty-five days.

It must not be supposed that the spots seen now upon the sun are the same spots that Fabricius saw so long ago. There is perpetual change going on; new spots forming, old spots vanishing; one spot breaking into two, two spots joining into one, and so on. Even in a single hour great alterations are sometimes seen to take place.

Still, many of the spots do remain long enough and keep their shapes closely enough to be watched from day to day, and to be known again as old friends when they reappear, after being about twelve days hidden on the other side of the sun. So that the turning of the sun upon his axis has become, after long and careful examination, a certain known fact.

For more than thirty years one astronomer kept close watch over the spots on every day that it was possible to see the sun. Much has been learned from his resolute perseverance.

Now what are these spots?

One thing seems pretty sure, and that is that they are caused by some kind of tremendous storms or cyclones taking place on the surface of that huge ball of fire.

The first attentive observer of the sun, Scheiner, at first regarded the spots as satellites—an indefensible opinion, which, however, some have attempted to revive. Galileo attributed them to clouds or vapors floating in the solar atmosphere; this was the best conclusion which could be drawn from the observations of that epoch. This opinion met for a long time with general approval; it has even been renewed in our day. Some astronomers, and among them

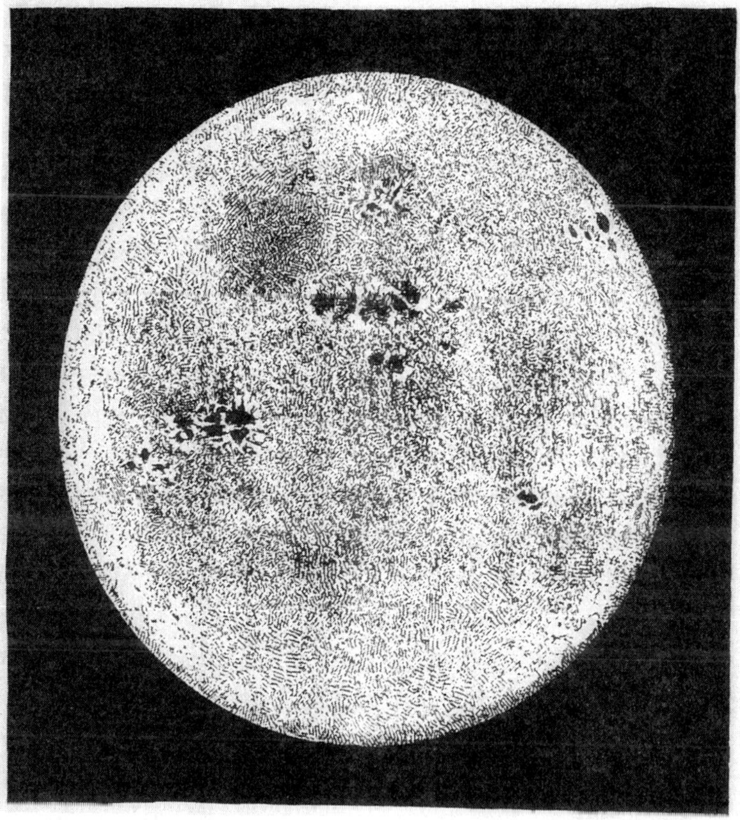

SUN SPOTS.

Lalande, believed, on the contrary, that they were mountains, of which the flanks, more or less steep, might produce the aspect of the penumbra—an opinion irreconcilable with the proper motion which the spots sometimes possess in a very marked manner. It is not usual, in fact, to see mountains traveling.

Derham attributed them to the smoke issuing from the volcanic craters of the sun, an opinion revived and maintained in recent times by Chacornac. Several *savants*, regarding the sun as a liquid and incandescent mass, have also explained the spots by immense cinders floating on this ocean of fire. But a century had scarcely elapsed after the epoch when the spots were observed for the first time, when an English astronomer, Wilson, showed with certainty that the spots are hollow.

We do not know with any certainty whether the sun is through and through one mass of glowing molten heat, or whether he may have a solid and even cool body within the blazing covering. Some have thought the one, and some the other. We only know that he is a mighty furnace of heat and flame, beyond anything that we can possibly imagine on our quiet little earth.

It seems very sure that no such thing as "quietness" is to be found on the surface of the sun. The wildest and fiercest turmoil of rushing wind and roaring flame there prevails. The cyclones or hurricanes which take place are sometimes so rapid and so tremendous in extent, tearing open the blazing envelope of the sun, and showing glimpses of fiery though darker depths below, that we, on our far-distant earth, can actually watch their progress.

Astronomers speak of the "quivering fringe of fire" all around the edge of the sun, visible through telescopes. But the sun is perpetually turning on his axis, so that each hour fresh portions of his surface thus pass the edge, showing the same appearance.

What conclusion can we come to but that the whole enormous surface of the sun is one restless, billowy sea of fire and flame?

It sounds to us both grand and startling to hear of the mighty outbreaks from Mount Vesuvius, or to read of glowing lava from a volcano in Hawaii pouring in one unbroken stream for miles.

But what shall be thought of rosy flames mounting to a height of fifty or a hundred thousand miles above the edge of the sun? What shall be thought of a tongue of fire long enough to fold three or four times round our solid earth? What shall be thought of the awful rush of burning gases, sometimes seen, borne along at the rate of one or two or even three hundred miles in a single second across the sun's surface? What shall be thought of the huge dark rents in this raging fiery ocean, rents commonly from fifty to one hundred thousand miles across, and not seldom more?

Fifty thousand miles! A mere speck, scarcely visible without a telescope; yet large enough to hold seven earths like ours flung in together. The largest spot measured was so enormous that eighteen earths might have been arranged in a row across the breadth of it, like huge boulders of rock in a mountain cavern; and to have filled up the entire hole about one hundred earths would have been needed.

It is well to grow familiar with certain names given by astronomers to certain parts of the sun.

The round shining disk or flat surface, seen by all of us, is called the *photosphere*, or "light-sphere." It has a tolerably well-defined edge or "limb," and dazzles the eye with its intense brightness.

Across the photosphere the spots move, sometimes many, sometimes few, in number. Besides the dark spots, there are spots of extreme brilliancy, standing out on even that dazzling surface, which causes a piece of white-hot iron to look black and cold by contrast. These extra-radiant spots which come and go, and at times change with great rapidity, are named *faculæ*—a Latin word meaning "torches."

At the edge of the photosphere astronomers see what has been already mentioned, that which is sometimes called the *chromotosphere*, which one has named "the *sierra*," and which another has described as "a quivering fringe of fire." The waves of the sea, on a stormy day, seen in the distance rising and breaking the horizon-line, may serve as an illustration; only in the sun the waves are of fire, not water. What must their height be, to be thus visible at a distance of ninety-three millions of miles?

Outside the *sierra* are seen, at certain seasons, bright, "rose-colored prominences" or flames of enormous height. During an eclipse, when the dark body of the moon comes between the sun and us, exactly covering the photosphere, these red tips stand out distinctly beyond the edge of sun and moon. Their changes have been watched, and their height repeatedly measured. Some are so lofty that ten little earths such as ours might be heaped up, one upon another, without reaching to their top. Whether they are in character at all like the outbursts from our own volcanoes it is hard to say. To compare the two would be rather like comparing a small kitchen fire with a mighty iron-smelting furnace. The faculæ, the sierra, and the prominences are visible only through a telescope.

Outside these red flames or "prominences" is the corona, commonly divided into the inner and outer coronas.

Many different explanations have been offered of this beautiful crown of light round the sun, plainly visible to the naked eye during an eclipse. But here again we know little, and must be content to watch and wait.

CHAPTER III.

WHAT BINDS THE FAMILY TOGETHER?

WHAT is it which binds together all the members of the Solar System? Ah, what? Why should not the sun at any moment rush away in one direction, the earth in a second, the planets in half a dozen others? What is there to hinder such a catastrophe? Nothing—except that they are all held together by a certain close family tie; or, more correctly, by the powerful influence of the head of the family.

This mysterious power which the sun has, and which all the planets have also in their smaller degrees, is called Attraction. Sometimes it is named Gravitation or Gravity. When we speak, as we often do, of the *law* of attraction or gravitation, we mean simply this—that throughout the universe, in things little and great, is found a certain wonderful *something*, in constant action, which we call a "law." What the "something" may be, man can not tell; for he knows it only by its effects. But these effects are seen everywhere, on all sides, in the earth and in the universe. It is well named in being called a "law;" for we are compelled to obey it. None but the Divine Lawgiver who made this law can for a single moment suspend its working.

What causes an apple to fall to the ground when it drops from the branch? Why should it not, instead, rise upwards? Because, of course, it is heavy, or has weight. But what *is* weight? Simply this—that the earth draws or drags everything downwards towards herself by the power of attraction. Every substance, great or small, light or heavy, is made up of tiny atoms. Each one of these atoms attracts or draws all the other atoms towards itself; and the closer they are together, the more strongly they pull one another.

The atoms in a piece of iron are much closer than the atoms in a piece of wood; therefore the iron is called the "more *dense*" of the two, and its weight or "mass" is greater. The more closely the atoms are pressed together, the greater the number of them in a small space, and the more strong the drawing towards the earth; for the earth draws each one of these atoms equally. That is only another way of saying that a thing is "heavier." If you drop a stone from the top of a cliff, will it rise upwards or float in the air? No, indeed. The pull of the earth's attraction, dragging and still dragging downward, makes it rush through the air, with speed quickening each instant, till it strikes the ground. Every single atom in every single body *pulls* every other atom, whether far or near. The nearer it is, the stronger always the pulling.

We do not always feel this, because the very much greater attraction of the earth hides—or smothers, as we may say—the lesser attraction of each small thing for another. But though you and I might stand side by side upon earth, and feel no mutual attraction, yet if we could mount up a few thousands of miles, far away from earth, and float in distant space, there we should find ourselves drawn together, and unable to remain apart.

Now precisely as an apple falling from a tree and a stone dropping from a cliff are dragged downward to the earth, just so our earth and all the planets are dragged downwards towards the sun and towards each other. The law of the earth's attraction of all objects on its surface to itself was indistinctly suspected a very long time ago; but it was the great Newton who first discovered that this same law was to be found working among the members of the whole Solar System. The sun attracts the earth, and the earth attracts the sun. But the enormous size of the sun compared with our earth—like a great nine-foot globe beside a tiny one-inch ball—makes our power of attraction to be quite lost sight of in his, which is so much greater.

The principle of gravitation is of far wider scope than we have yet indicated. We have spoken merely of the attraction of the earth, and we have stated that its attraction extends throughout space. But the law of gravitation is not so limited. Not only does the earth attract every other body, and every other body attract the earth, but each of these bodies attracts each other; so that, in its more complete shape, the law of gravitation announces that "every body in the universe attracts every other body with a force which varies inversely as the square of the distance." It is impossible for us to overestimate the importance of this law. It supplies the clue by which we can unravel the complicated movements of the planets. It has led to marvelous discoveries, in which the law of gravitation has enabled us to anticipate the telescope, and, indeed, actually to feel the existence of bodies before those bodies have even been seen.

We come now to another question. If the sun is pulling with such power at the earth and all her sister planets, why do they not fall down upon him? What is to prevent their rolling some day into one of those deep rents in his fiery envelope? Did you ever tie a ball to a string, and swing it rapidly round and round your head? If you did, you must have noticed the steady outward pull of the ball. The heavier the ball, and the more rapid its whirl, the stronger the pull will be. Let the string slip, and the rush of the ball through the air to the side of the room will make this yet more plain. Did you ever carry a glass of water quickly along, and then, on suddenly turning a corner, find that the water has not turned with you? It has gone on in its former direction, leaving the glass, and spilling itself on the floor.

The cause in both cases is the same. Here is another "law of nature," so-called. Though we can neither explain nor understand why and how it is so, we see it to be one of the fixed rules of God's working in every-day life throughout his universe. The law, as we see it, seems to be this: Everything which is at rest must remain at rest, until set moving by some cause outside of, or independent of itself; and everything which is once set moving, must continue moving in a straight line until checked.

According to this a cannon-ball lying on the ground ought to remain there until it is set in motion; and, once set in motion by being fired from a cannon, it ought to go on forever. Exactly so—if nothing stops it. But the earth's attraction draws the cannon-ball downward, and every time it strikes the ground it is partly checked. Also each particle of air that touches it helps to bring it to rest. If there were no earth and no air in the question, the cannon-ball might rush on in space for thousands of years.

Why did the water get spilled? Because it necessarily continued moving in a straight line. Your sudden change of direction compelled the solid glass to make the same change, but the liquid water was free to go straight on in its former course; so it obeyed this law, and *did* go on. Why did the ball pull hard at the string as you swung it round? Because at each instant it was striving to obey this same law, and to rush onward in a straight line. The pull of the string was every moment fighting against that inclination, and forcing the ball to move in a circle.

Just such is the earth's movement in her yearly journey round the sun. The string holding in the ball pictures the sun's attraction holding in the earth. The pulling of the ball outward in order to continue its course in a straight line, pictures the pulling of our earth each moment to break loose from the sun's attraction and to flee away into distant space.

For the earth is not at rest. Each tick of the clock she has sped onward over more than eighteen miles of her pathway through the sky. Every instant the sun is dragging, with the tremendous force of his attraction, to make her fall nearer to him. Every instant the earth is dragging with the tremendous force of her rapid rush, to get away from him. These two pullings so far balance each other, or, more strictly, so far combine together, that between the two she journeys steadily round and round in her nearly circular orbit.

If the sun pulled a little harder she would need to travel a little faster, or she would gradually go nearer to him. If the earth went faster, and the sun's attraction remained the same as it is now, she would gradually widen her distance. Indeed, it would only be needful for the earth to quicken her pace to about five-and-twenty miles a second, the sun's power to draw her being unchanged, and she would then wander away from him for ever. Day by day we on our earth should travel farther and farther away, leaving behind us all light, all heat, all life, and finding ourselves slowly lost in darkness, cold, and death.

For what should we do without the sun? All our light, all our warmth, come from him. Without the sun, life could not exist on the earth. Plants, herbage, trees, would wither; the waters of rivers, lakes, oceans, would turn to masses of ice; animals and men would die. Our earth would soon be one vast, cold, forsaken tomb of darkness and desolation.

We may not think so, but everything which moves, circulates, and lives on our planet is the child of the sun. The most nutritious foods come from the sun. The wood which warms us in winter is, again, the sun in fragments. Every cubic inch, every pound of wood, is formed by the power of the sun. The mill which turns under the impulse of wind or water, revolves only by the sun. And in the black night, under the rain or snow, the blind and noisy train, which darts like a flying serpent through the fields, rushes along above the valleys, is swallowed up under the mountains, goes hissing past the stations, of which the pale eyes strike silently through the mist,—in the midst of night and cold, this modern animal, produced by human industry, is still a child of the sun. The coal from the earth which feeds its stomach is solar work, stored up during millions of years in the geological strata of the globe.

As it is certain that the force which sets the watch in motion is derived from the hand which has wound it, so it is certain that all terrestrial power proceeds from the sun. It is its heat which maintains the three states of bodies—solid, liquid, and gaseous. The last two would vanish, there would be nothing but solids; water and air itself would be in massive blocks,—if the solar heat did not maintain them in the fluid state. It is the sun which blows in the air, which flows in the water, which moans in the tempest, which sings in the unwearied throat of the nightingale. It attaches to the sides of the mountains the sources of the rivers and glaciers, and consequently the cataracts and the avalanches are precipitated with an energy which they draw directly from him. Thunder and lightning are in their turn a manifestation of his power. Every fire which burns and every flame which shines has received its life from the sun. And when two armies are hurled together with a crash, each charge of cavalry, each shock between two army corps, is nothing else but the misuse of mechanical force from the same star. The sun comes to us in the form of heat, he leaves us in the form of heat; but between his arrival and his departure, he has given birth to the varied powers of our globe.

I have spoken before about the old-world notion that our earth was a fixed plain, with the sun circling round her. When the truth dawned slowly upon some great minds, anxious only to know what really was the truth, others made a hard struggle for the older and pleasanter mode of thinking. It went with many sorely against the grain to give up all idea of the earth being the chief place in the universe. Also there was something bewildering and dizzying in the notion that our solid world is never for one moment still. But truth won the victory at last. Men consented slowly to give up the past dream, and to learn the new lesson put before them.

We still talk of the sun rising and setting, and of the stars doing the same. This is, however, merely a common form of speech, which means just the opposite. For instead of the sun and stars moving, it is the earth which moves.

The earth has two distinct movements. Indeed, I ought to say that she has three; but we will leave all thought of the third for the present. First, she turns round upon her axis once in every twenty-four hours. Secondly, she travels round the sun once in about every three hundred and sixty-five days and a quarter. No wonder our ancestors were startled to learn that the world, which they had counted so immovable, was perpetually spinning like a humming-top, and rushing through space like an arrow.

You may gain some clear notions as to the daily rising and setting of our sun, with the help of an orange. Pass a slender knitting-needle through the orange, from end to end, and hold it about a yard distant from a single candle in a room otherwise darkened. Let the needle or axis slant somewhat, and turn the orange slowly upon it.

The candle does not move; but as the orange turns, the candle-light falls in sucession upon each portion of the yellow rind. Half of the orange is always in shade, and half is always in light; while at either side, if a small fly were standing there, he would be passing round out of shade into candle-light, or out of candle-light into shade.

Each spot on our earth moves round in turn into half-light, full-light, half-light, and darkness; or, in other words, has morning-dawn, midday-light, evening twilight, and night. Each spot on our earth would undergo regularly these changes every twenty-four hours throughout the year, were it not for another arrangement which so far affects this that the North and South Poles are, by turns, cut off from the light during many months together.

Thus the sun is in the center of the Solar System, turning slowly on his axis; and the earth and the planets travel round him, each spinning like a teetotum, so as to make the most of his bright warm rays. But for this spinning movement of the earth, our day and night, instead of being each a few hours long, would each last six months.

You may notice that, as you turn the orange steadily round, the outside surface of the skin has to move much more slowly in those parts close to the knitting-needle, than in those parts which bulge out farthest from it. Near the North and South Poles the surface of our earth travels slowly round a very small circle in the course of twenty-four hours. But at the equator every piece of ground has to travel about twenty-five thousand miles in the same time; so that it rushes along at the rate of more than one thousand miles an hour. A man standing on the earth at the equator is being carried along at this great speed, not *through* the air, for the whole atmosphere partakes of the same rapid motion, but *with* the air, round and round the earth's axis.

Now about the other movement of the earth—her yearly journey round the sun. While she moves, the sun, as seen from the earth, seems to change his place. First he is observed against a background of one group of stars, then against a second, then against a third. Not that the stars are visible in the daytime when the sun is shining, but their places are well known in the heavens; and also they can be noted very soon after he sets, or before he rises, so that the constellations nearest to him may each day be easily found out.

Of course, in old times the sun was thought to be really taking this journey among the stars, and men talked of "the sun's path" in the heavens. This path was named "the Ecliptic," and we use the word still, though we know well that the movements are not really his, but ours.

As the earth's daily movement causes day and night, so the earth's yearly movement causes spring, summer, autumn, and winter. A few pages back I mentioned, in passing, one slight yet important fact which lies at the root of this matter about the seasons. The earth, journeying round the sun, travels with her axis *slanting*. Put your candle in the middle of the table, and stand at one end, holding your orange. Now let the knitting-needle, with the orange upon it, so slant that one end shall point straight over the candle, towards the upper part of the wall at the farther end of the room. Call the upper end so pointing the North Pole of your orange. You will see that the candle-light falls chiefly upon the upper half of the orange; and as you turn it slowly, to picture day and night, you will find that the North Pole has no night, and the South Pole has no day. That is summer in the northern hemisphere, and winter in the southern.

Walk round next to one side of the table, towards the right hand, taking care to let the knitting-needle point steadily still in exactly the same direction, not towards the same *spot*, for that would alter its direction as you move, but towards the same *wall*. Stop, and you will find the candle lighting up one-half of your orange, from the North to the South Poles. Turn it slowly, never altering the slope of the axis, and you will see that every part of the orange comes by turns under the light. This is the Autumnal Equinox, when days and nights all over the world are equal in length.

Walk on to the other end of the table, still letting the needle slope and point steadily as before. Now the candle-light will shine upon the lower or South Pole, and the North Pole will be entirely in the shade. This is summer in the southern hemisphere, and winter in the northern.

Pass on to the fourth side of the table, and once

more you will find it, as at the second side, equal light from North Pole to South Pole. This will be the Spring Equinox.

It is an illustration that may be easily practiced. But everything depends upon keeping the slant of the needle or axis unchanged throughout. If it be allowed to point first to right, and then to left, first towards the ceiling, and then towards the wall, the attempt will prove a failure.

CHAPTER IV.

THE LEADING MEMBERS OF OUR FAMILY—FIRST GROUP.

THE chief distinction between stars and planets is, as before said, that the stars shine entirely by their own light, while the planets shine chiefly, if not entirely, by reflected light. The stars are suns; mighty globes of glowing flame. The planets simply receive the light of the sun, and shine with a brightness not their own.

A lamp shines by its own light; but a looking-glass, set in the sun's rays and flashing beams in all directions, shines by reflected light. In a dark room it would be dark. If there were no sun to shine upon Mars or Venus, we should see no brightness in them. The moon is like the planets in this. She has only borrowed light to give, and none of her own.

Any one of the planets removed to the distance of the nearest fixed star, would be invisible to us. Reflected light will not shine nearly so far as the direct light of a burning body. There may be thousands or millions of planets circling round the stars—those great and distant suns—just as our brother-planets circle round our sun. But it is impossible for us to see them. The planets which we can see are close neighbors, compared with the stars. I do not mean that they are near in the sense in which we speak of nearness upon earth. They are only near in comparison with what is so very much farther away.

For a while we must now leave alone all thought of the distant stars, and try to gain a clear idea of the chief members of our own Family Circle—that family circle of which the sun is the head, the center, the source of life and warmth and light. There are two ways in which astronomers group the planets of the Solar System. One way is to divide them into the Inferior Planets, and the Superior Planets.

As the earth travels in her pathway round the sun, two planets travel on their pathways round the sun nearer to him than ourselves. If the pathway or orbit of our earth were pictured by a hoop laid upon the table, with a ball in the center for the sun; then those two planets would have two smaller hoops of different sizes *within* ours; and the rest would have larger hoops of different sizes *outside* ours. The two within are called inferior planets, and the rest outside are called superior planets.

A round hoop would not make a good picture of an orbit. For the yearly pathway of our earth is not in shape perfectly round, but slightly oval; and the sun is not exactly in the center, but a little to one side of the center. This is more or less the case with the orbits of all the planets.

But the laying of the hoops upon the table would give no bad idea of the way in which the orbits really lie in the heavens. The orbits of all the chief planets do not slope and slant round the sun in all manner of directions. They are placed almost in the same *plane* as it is called—or, as we might say, in the same *flat.* In these orbits the planets all travel round in the same direction. One may overtake a second on a neighboring orbit, and get ahead of him, but one planet never goes back to meet another.

In speaking of the orbits, I do not mean that the planets have visible marked pathways through the heavens, any more than a swallow has a visible pathway through the sky, or a ship a marked pathway through the sea. Yet each planet has his own orbit, and each planet so distinctly keeps to his own, that astronomers can tell us precisely whereabouts in the heavens any particular planet will be, at any particular time, long years beforehand.

There is also another mode of grouping the planets, besides dividing them into superior and inferior planets. By this other mode we find two principal groups or quartets of planets, separated by a zone or belt of a great many very small planets.

First Group........ { Mercury. Venus. Earth. Mars.

The Asteroids or Planetoids.

Second Group...... { Jupiter. Saturn. Uranus. Neptune.

The first four are small compared with the last four, though much larger than any in the belt of tiny Asteroids.

It was believed at one time that a planet had been discovered nearer to the sun than Mercury, and the name Vulcan was given to it. But no more has been seen of Vulcan, and his existence is so doubtful that we must not count him as a member of the family without further information.

Mercury is very much smaller than our Earth. The diameter of the earth is eight thousand miles, but the diameter of Mercury is only about three thousand miles—not even half that of the earth. Being so much nearer to the sun than ourselves, the pulling of his attraction is much greater, and this has to be balanced by greater speed, or Mercury would soon fall down upon the sun. Our distance from the sun is ninety-three millions of miles. Mercury's distance is only about one-third of ours; and instead of traveling, like the earth, at the rate of eighteen miles each second, Mercury dashes headlong through space at the mad pace of twenty-nine miles each second. It is a good thing the earth does not follow his example, or she would soon break loose from the sun's control altogether.

The earth takes more than three hundred and sixty-five days, or twelve months, to journey round the sun in her orbit. That is what we call "the length of our year." But Mercury's year is only eighty-eight days, or not quite three of our months. No wonder!—when his pathway is so much shorter, and his speed so much greater than ours. So Mercury has four years to one year on earth; and a person who had lived on Mercury as long as five earthly years, would then be twenty years old. The increased number of birthdays would scarcely be welcome in large families, supposing we could pay a long visit there.

The sun, as seen from Mercury, looks about four and a half times as large as from here; the heat and glare being increased in proportion. No moon has ever been found, belonging to Mercury.

The planet Mercury is, like the earth and the moon, a globe of dark matter which only shines and is visible by the illumination of the solar light. Its motion round the central star, which brings it sometimes between the sun and us, sometimes in an oblique direction, sometimes at right angles, and shows us a part incessantly variable, of its illuminated hemisphere, produces in its aspect, as seen in a telescope, a succession of phases similar to those which the moon

PHASES OF MERCURY BEFORE INFERIOR CONJUNCTION—EVENING STAR.

presents to us. The cut represent the apparent variations of size and the succession of phases, visible in the evening after sunset; when the planet attains its most slender crescent, it is in the region of its orbit nearest to the earth, and passes between the sun and us; then, some weeks afterwards, it emerges from the solar rays and passes again through the same series of phases in inverse order, as we see them by reversing the figure.

Venus, the second inferior planet, is nearly the same size as our earth. Seen from the earth, she is one of the most brilliant and beautiful of all the planets.

Her speed is three miles a second faster than ours, and her distance from the sun is about two-thirds that of our own; so that the orbit of Venus lies half-way between the orbit of Mercury and the orbit of the earth. The day of Venus is about half an hour shorter than ours. Her year is nearly two hundred and twenty-five days, or seven and a half of our months. One or two astronomers have fancied that they caught glimpses of a moon near Venus; but this is still quite doubtful, and indeed it is believed to have been a mistake.

Venus and Mercury are only visible as morning and evening planets. Venus, being farther from the sun, does not go before and follow after him quite so closely as Mercury, and she is therefore the longer within sight.

When Venus, traveling on her orbit, comes just between the sun and us, her dark side is turned towards the earth, and we can catch no glimpse of her. When she reaches that part of her orbit which is farthest from us, quite on the other side of the sun, her great distance from us makes her light seem less. But about half-way round on either side, she shows exceeding brilliancy, and that is the best view we can get of her.

Seen through a telescope, Venus undergoes phases like those of Mercury, or of our moon. That is to say, we really have "new Venus," "quarter Venus," "half Venus," "full Venus," and so on.

Of all the luminaries in the heavens, the sun and moon excepted, the planet Venus is the most conspicuous and splendid. She appears like a brilliant lamp amid the lesser orbs of night, and alternately anticipates the morning dawn, and ushers in the evening twilight. When she is to the westward of the sun, in winter, she cheers our mornings with her vivid light, and is a prelude to the near approach of the break of day and the rising sun. When she is eastward of that luminary, her light bursts upon us after sunset, before any of the other radiant orbs of heaven make their appearance; and she discharges, in some measure, the functions of the absent moon. The brilliancy of this planet has been noticed in all ages, and has been frequently the subject of description and admiration both by shepherds and by poets. The Greek poets distinguished it by the name of Phosphor, "light-bringer," when it rose before the sun, and Hesperus, "the west," when it appeared in the evening after the sun retired. It is now generally distinguished by the name of the Morning and Evening Star.

Next to the orbit of Venus comes the orbit of our own Earth, the third planet of the first group.

Mars, the fourth of the inner quartet, but the first of the superior planets, is a good deal smaller than Venus or the earth. The name Mars, from the heathen god of war, was given on account of his fiery reddish color. Mars is better placed than Venus for being observed from earth. When he is at the nearest point of his orbit to us, we see him full in the blaze of sunlight; whereas Venus, at her nearest point, turns her bright face away.

The length of the day of Mars—or, in other words, the time he takes to turn upon his axis—is only forty minutes longer than that of earth. Mars' journey round the sun is completed in the course of six hundred and eighty-seven days, not much less than two of our years. His distance from the sun is about one hundred and forty millions of miles, and his speed is fourteen miles a second. We shall find, with the increasing distance of each planet, that the slower pace balances the lessened amount of the sun's attraction.

Passing on from Mars, the last of the first group of planets, we reach the belt of Asteroids, sometimes called Planetoids, Minor Planets, or Telescopic Planets. They are so tiny that Mercury is a giant compared with the largest among them.

The zone of space containing all these little planets is more than a hundred millions of miles broad. Their orbits do not lie flat in almost the same plane, but slant about variously in a very entangled fashion. If a neat model were made of this zone, with a slender piece of wire to represent each orbit, it would be found impossible to lift up one wire without pulling up all the rest with it. Those asteroids lying nearest to the sun take about three of our years to travel round him, and those lying farthest take about six of our years.

New members of the group are very often found. The number of asteroids now known amounts to about three hundred and twenty. Ceres is estimated by Professor Barnard, of the Yerkes Observatory, Chicago University, to be six hundred miles in diameter. Vesta is about two hundred and fifty miles in diameter. Meta, on the other hand, is less than seventy-five miles. There are some smaller ones that do not measure twenty miles in diameter, and it is probable that there are many which are so small as to be absolutely invisible to the best telescopes, and which measure only a few hundred rods in diameter. Twenty thousand Vestas would be needed to make one globe equal to our earth in size.

Are they *worlds?* Why not? Is not a drop of water, shown in the microscope, peopled with a multitude of various beings? Does not a stone in a meadow hide a world of swarming insects? Is not the leaf of a plant a world for the species which inhabit and prey upon it? Doubtless among the multitude of small planets there are those which must remain desert and sterile, because the conditions of life (of any kind) are not found united. But we can not doubt that on the majority the ever-active forces of nature have produced, as in our world, creations appropriate to these minute planets. Let us repeat, moreover, that for nature there is neither great nor little. And there is no necessity to flatter ourselves with a supreme disdain for these little worlds; for in reality the inhabitants of Jupiter would have more right to despise us than we have to despise Vesta, Ceres, Pallas, or Juno. The disparity is greater between Jupiter and the earth than between the earth and these planets. A world of two, three, or four hundred miles in diameter is still a continent worthy to satisfy the ambition of a Xerxes or a Tamerlane.

CHAPTER V.

THE LEADING MEMBERS OF OUR FAMILY—SECOND GROUP.

LEAVING behind us the busy zone of planetoids, hurrying round and round the sun in company, we cross a wide gap, and come upon a very different sight.

The distance from the sun which we have now reached is little less than four hundred and fifty millions of miles, or about five times as much as the earth's distance; and the sun in the heavens shows a diameter only one-fifth of that which we are accustomed to see. Slowly—yet not slowly—floating onwards through space, in his far-off orbit, we find the magnificent planet Jupiter.

Is eight miles each second slow progress? Compared with the wild whirl of little Mercury, or even compared with the rate of our own earth's advance, we may count it so; but certainly not, compared with our notions of speed upon earth. Eight miles each second is five hundred times as fast as the swiftest express-train ever made by man. No mean pace that for so enormous a body. For Jupiter is the very largest of all the members of the Solar System except the sun himself—quite the eldest brother of the family. His diameter is about eighty-five thousand miles, his axis being nearly eleven times as long as that of earth. Though in proportion to his great bulk not nearly so heavy as our earth, yet his bulk is so vast that more than twelve hundred earths would be needed to make one Jupiter.

It must not, however, be forgotten that there is a certain amount of uncertainty about these measurements of Jupiter. He seems to be so covered with a dense atmosphere and heavy clouds that it is quite impossible for us to learn the exact size of the solid body within.

Jupiter does not travel alone. Borne onwards with him, and circling steadily around him, are five moons; the smallest with a probable diameter of about one hundred miles; one about the same size as our own moon, and the others all larger. The nearest of the five, though distant from the center of the planet over one hundred and twelve thousand miles, revolves about him in twelve hours; the second, though farther from Jupiter than our moon from the earth, speeds round him in less than two of our days. The most distant, though over a million miles away, takes scarcely seventeen days to accomplish its long journey. Jupiter and his moons make a little system by themselves—a family circle within a family circle.

Like the smaller planets, Jupiter spins upon his axis; and he does this so rapidly that, notwithstanding his great size, his day lasts only ten hours, instead of twenty-four hours like ours. But if Jupiter's day is short, his year is not. Nearly twelve of our years pass by before Jupiter has traveled once completely round the sun. So a native of earth who had just reached his thirty-seventh year, would, on Jupiter, be only three years old.

Passing onward from Jupiter, ever farther and farther from the sun, we leave behind us another vast and empty space—empty as we count emptiness, though it may be that there is in reality no such thing as emptiness throughout the length and breadth of the universe. The width of the gap which divides the pathway of Jupiter from the pathway of his giant brother-planet Saturn is nearly five times as much as the width of the gap separating the earth from the sun. The distance of Saturn from the sun is not much less than double the distance of Jupiter.

With this great space in our rear, we come upon another large and radiant planet, the center, like Jupiter, of another little system; though it can only be called "little" in comparison with the much greater Solar System of which it forms a part.

Saturn's diameter is less than that of Jupiter, but the two come near enough to be naturally ranked together. Nearly seven hundred earths would be needed to make one globe as large as Saturn. But here again the dense and cloudy envelope makes us very uncertain about the planet's actual size. Saturn is like Jupiter in being made of lighter materials than our earth; and also in his rapid whirl upon his axis, the length of his "day" being only ten and a half of our hours.

From Jupiter's speed of eight miles each second, we come down in the case of Saturn to only five miles each second. And Jupiter's long annual journey looks almost short, seen beside Saturn's long journey of thirty earthly years. A man aged sixty, according to our fashion of reckoning time, would on Saturn have just kept his second birthday.

The system or family of Saturn is yet more wonderful than that of Jupiter. Not five only but eight moons travel ceaselessly round Saturn, each in its own orbit; and in addition to the eight moons, he has revolving round him three magnificent rings. These rings, as well as the moons, shine, not by their own brilliancy, for they have none, but by borrowed sunlight. The farthest of the moons wanders in his lonely pathway about two millions of miles away from Saturn. The largest of them is believed to be about the same size as the planet Mars. Of the three rings circling round Saturn, almost exactly over his equator, the inside one is dusky, purplish, and transparent; the one outside or over that is very brilliant; and the third, outside the second, is rather grayish in hue.

Another vast gap—more enormous than the last. It is a wearisome journey. From the orbit of Jupiter to the orbit of Saturn at their nearest points, was five times as much as from the sun to the earth. But from the orbit of Saturn to the orbit of Uranus, the next member of the sun's family, we have double even that great space to cross.

Still, obedient to the pulling of the sun's attractive power, Uranus wanders onward in his wide pathway round the sun, at the rate of four miles a second. Eighty-four of our years make one year of Uranus. This planet has four moons, and thus forms a third smaller system within the Solar System; but he may have other satellites also, as yet undiscovered. In size he is seventy-four times as large as our earth.

One more mighty chasm of nine hundred millions of miles, for the same distance which separates the pathway of Saturn from the pathway of Uranus, separates also the pathway of Uranus from the pathway of Neptune. Cold, and dark, and dreary indeed seems to us the orbit on which this banished member of our family circle creeps round the sun, in the course of one hundred and sixty-five years, at the sluggish rate of three miles a second.

On the planet Saturn, the quantity of light and heat received from the sun is not much more than a hundredth part of that which we are accustomed to receive on earth. But by the time we reach Neptune, the great sun has faded and shrunk in the distance until to our eyes he looks only like an exceedingly brilliant and dazzling star.

We know little of this far-off brother, Neptune, except that he is rather larger than Uranus, being one hundred and five times as big as the earth; that he has at least one moon; and also that, like Uranus, he is made of materials lighter than those of earth, but heavier than those of Jupiter or Saturn.

After all, it is no easy matter to gain clear ideas as to sizes and distances from mere statements of "so many miles in diameter," and "so many millions of miles away." A "million miles" carries to the mind a very dim notion of the actual reality.

Now if we can in imagination bring down all the members of the Solar System to a small size, keeping always the same proportions, we may find it a help. "Keeping the same proportions" means that all must be lessened alike, all must be altered in the same degree. Whatever the supposed size of the earth may be, Venus must be still about the same size as the earth, Saturn seven hundred times as large, and so on. Also, whatever the distance of the earth from the sun, in miles, or yards, or inches, Mercury must still be one-third as far, Jupiter still five times as far, and thus with the rest.

First, as to size alone. Suppose the earth is represented by a small globe, exactly three inches in diameter. It will be a very small globe. Not only men and houses, but mountains, valleys, seas, will all have to be reduced to so minute a size, as to be quite invisible to the naked eye.

Fairly to picture the other members of the Solar System, in due proportion, you will have them as follows:

Mercury and Mars will be balls smaller than the earth, and Venus nearly the same size of the earth. Uranus and Neptune will be each somewhere about a foot in diameter. Saturn will be twenty-eight inches and Jupiter thirty-two inches in diameter. The sun will be a huge dazzling globe, *twenty-six feet* in diameter. No wonder he weighs seven hundred and fifty times as much as all his planets put together.

Next let us picture the system more exactly on another and smaller scale. First, think of the sun as a brilliant globe, about nine feet, or three yards, in diameter, floating in space.

About one hundred yards from the sun travels a tiny ball, not half an inch in diameter, passing slowly round the sun—slowly, because as sizes and distances are lessened, speed must in due proportion be lessened also. This is Mercury. About two hundred yards from the sun travels another tiny ball, one inch in diameter. This is Venus. Nearly a quarter of a mile from the sun travels a third tiny ball, one inch again in diameter; and at a distance of two feet and a half from it a still smaller ball, one quarter of an inch in diameter, journeys round it and with it. These are the earth and the moon. About half as far again as the last-named ball, travels another, over half an inch in diameter. This is Mars.

Then comes a wide blank space, followed by a large number of minute objects, no bigger than grains of powder, floating round the sun in company. These are the Asteroids. Another wide blank space succeeds the outermost of them.

About one mile distant from the sun journeys a globe, ten inches in diameter. Round him, as he journeys, there travel five smaller balls, the largest of which is about the third of an inch in diameter. Their distances from the bigger globe vary from one and a third to twelve and a half feet. These are Jupiter and his moons.

Nearly two miles distant from the sun journeys another globe, about eight and a half inches in diameter. Eight tiny balls and three delicate rings circle round him as he moves. These are Saturn and his belongings. About four miles distant from the sun journeys another globe, four inches in diameter, with four tiny balls accompanying him—Uranus and his moons. Lastly, at a distance of six miles from the sun, one more globe, not much larger than the last, with one tiny companion, pursues his far-off pathway.

These proportions as to size and distance will serve to give a clear idea of the Solar System.

CHAPTER VI.

THE MOON.

Come, and let us pay a visit to the moon. We seem to feel a personal interest in her, just because she is, in so peculiar a sense, our own friend and close attendant. The sun shines for us; but, then, he shines for all the members of the Solar System. And the stars—so many as we can see of them—shine for us too; but no doubt they shine far more brilliantly for other and nearer worlds. The moon alone seems to belong especially to ourselves.

Indeed, we are quite in the habit of speaking about her as "our moon." Rather a cold and calm friend, some may think her, sailing always serenely past, whatever may be going on beneath her beams; yet she has certainly proved herself constant and faithful in her attachment.

We have not very far to travel before reaching her, —merely about two hundred and forty thousand miles. That is nothing, compared with the weary millions of miles which we have had to cross to visit some members of our family. A rope two hundred and forty thousand miles long would fold nearly ten times round the earth at the equator. You know the earth's diameter—about eight thousand miles. If you had thirty poles, each eight thousand miles long, and could fasten them all together, end to end, one beyond another, you would have a rod long enough to reach from the earth to the moon.

Let us take a good look at her before starting. She is very beautiful. That soft silvery light, so unlike sunlight or gaslight, or any other kind of light seen upon earth, has made her the darling of poets and the delight of all who love nature. Little children like to watch her curious markings, and to make out

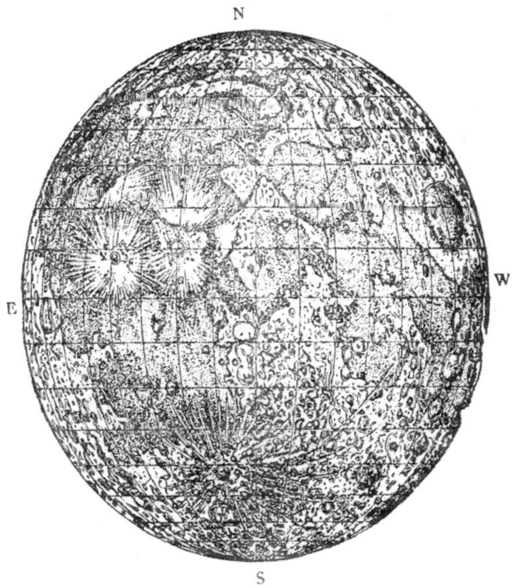

APPEARANCE OF THE FULL MOON.

the old man with his bundle of sticks, or the eyes, nose, and mouth of the moon—not dreaming what those markings really are. And in moods of sadness, how the pure calm moonlight seems to soothe the feelings! Who would suppose that the moon's beauty is the beauty rather of death than of life?

The stars have not much chance of shining through her bright rays. It is well for astronomers that she is not always at the full. But when she is, how large she looks—quite as large as the sun, though in reality her size, compared with his, is only as a very small pin's head compared with a school globe two feet in diameter. Her diameter is little more than two thousand miles, or one quatrer that of our earth; and her whole surface, spread out flat, would scarcely equal North and South America, without any of the surrounding islands.

The reason she looks the same size as the sun, is that she is so very much nearer. The sun's distance from us is more than one-third as many *millions* of miles as the moon's distance is *thousands* of miles. This makes an enormous difference.

We call our friend a "moon," and say that she journeys round the earth, while the earth journeys round the sun. This is true, but it is only part of the truth. Just as certainly as the earth travels round the sun, so the moon also travels round the sun. And just as surely as the earth is a planet, so the moon also is a planet. It is a common mode of expression to talk about "the earth and her satellite." A no less correct, if not more correct, way would be to talk of ourselves as "a pair of planets," journeying round the same sun, each pulled strongly towards him, and each pulling the other with a greater or less attraction, according to her size and weight. For the sun actually does draw the moon with more force than that with which the earth draws her. Only as he draws the earth with the same sort of force, and nearly in the same degree, he does not pull them apart.

The moon, like the other planets, turns upon her axis. She does this very slowly, however; and most singularly she takes exactly the same time to turn once upon her axis that she does to travel once round the earth. The result of this is, that we only see one face of the moon. If she turned upon her axis, and journeyed round the earth in two different lengths of time, or if she journeyed round us and did not turn upon her axis at all, we should have views of her on all sides, as of other planets. But as her two movements so curiously agree, it happens that we always have one side of the moon towards us, and never catch a glimpse of the other side.

And now we are ready to start on our journey of two hundred and forty thousand miles. An express train, moving ceaselessly onward night and day, at the rate of sixty miles an hour, would take us there in about five months and a half. But no line of rails has ever yet been laid from the earth to the moon, and no "Flying Dutchman" has ever yet plied its way to and fro on that path through the heavens. Not on the wings of steam, but on the wings of imagination, we must rise aloft. Come—it will not take us long. We shall pass no planets or stars on the road, for the moon lies nearer to us than any other of the larger heavenly bodies.

Far, far behind us lies the earth, and beneath our feet, as we descend, stretch the broad tracts of moonland. For "downward" now means towards the moon, and away from the distant earth.

What a strange place we have reached! The weird, ghastly stillness of all around, and the glaring, dazzling, cloudless heat, strike us first and most forcibly. Nothing like this heat have we ever felt on earth. For the close of the moon's long day—on this side of its globe—is approaching, and during a whole fortnight past the sun's fierce rays have been beating down on these shelterless plains. Talk of sweltering tropics on earth! What do you call *this* heat? Look at the thermometer: how the quicksilver is rising! Well, that is not a common thermometer which we have brought with us. The mercury has stopped—three hundred degrees above boiling water!

Not a cloud to be seen overhead; only a sky of inky blackness, with a blazing sun, and thousands of brilliant stars, and the dark body of our own earth, large and motionless, and rimmed with light. Seen from earth, sun and moon look much the same size; but seen from the moon, the earth looks thirteen times as large as our full moon. Not even a little mistiness in the air to soften this fearful glare! Air! why, there is no air; at least not enough for any human being to breathe or feel. If there were air, the sky would be blue, not black, and the stars would be invisible in the daytime. It looks strange to see them now shining beside the sun.

And then, this deadly stillness! Not a sound, not a voice, not a murmur of breeze or water. How could there be? Sound can not be carried without air, and of air there is none. As for breeze—wind is moving air, and where we have no air we can have no wind. As for water—if there ever was any water on the moon, it has entirely disappeared. We shall walk to and fro vainly in search of it now. No rivers, no rills, no torrents in those stern mountain ramparts rising on every side. All is craggy, motionless, desolate.

How very, very slowly the sun creeps over the black

sky! And no marvel, since a fortnight of earth-time is here but one day, answering to twelve hours upon earth. Can not we find shelter somewhere from this blazing heat? Yonder tall rock will do, casting a sharp shadow of intense blackness. We never saw such shadows upon earth. There the atmosphere so breaks and bends and scatters about the light, that outlines of shadows are soft and hazy, even the clearest and darkest of them, compared with this.

When will the sun go down? But he is well worth looking at meanwhile. How magnificent he appears, with his pure radiant photosphere, fringed by a sierra of dazzling pink and white, orange and gold, purple and blue. For here no atmosphere lies between to blend all into yellow-white brightness. And how plainly stand out those prominences or tongues of tinted flame, not merely rose-colored, as seen from earth, but matching the sierra in varying hues; while beyond spreads a gorgeous belt of pink and green, bounded by lines and streams of delicate white light, reaching far and dying slowly out against the jet background. The black spots on the face of the sun are very distinct, and so also are the brilliant faculæ.

We must take a look around us now at moonland, and not only sit gazing at the sun, though such a sky may well enchain attention. How unlike our earthly landscapes! No sea, no rivers, no lakes, no streams, no brooks, no trees, bushes, plants, grass, or flowers; no wind or breeze; no cloud or mist or thought of possible rain; no sound of bird or insect, of rustling leaves or trickling water. Nothing but burning, steadfast, changeless glare, contrasting with inky shadows;

sun and earth and stars in a black heaven above; silent, desolate mountains and plains below.

For though we stand here upon a rough plain, this moon is a mountainous world. Ranges of rugged hills stretch away in the distance, with valleys lying between—not soft, green, sloping, earthly valleys, but steep gorges and precipitous hollows, all white dazzle and deep shade.

But the mountains do not commonly lie in long ranges, as on earth. The surface of the moon seems to be dented with strange round pits, or craters, of every imaginable size. We had a bird's-eye view of them as we descended at the end of our long journey moonward. In many parts the ground appears to be quite honeycombed with them. Here are small ones near at hand, and larger ones in the distance. The smaller craters are surrounded by steep ramparts of rock, the larger ones by circular mountain-ranges. We have nothing quite like them on earth.

Are they volcanoes? So it would seem; only no life, no fire, no action, remain now. All is dead, motionless, still. Is this verily a blasted world? Has it fallen under the breath of Almighty wrath, coming out scorched and seared? Is it simply passing through a certain burnt-out, chilled phase of existence, through which other planets also pass, or will pass, at some stage of their career? Who can tell?

We will move onward, and look more closely at that towering mass of rugged rocks, beyond which the sun will by and by go down. Long jetty shadows lie from them in this direction. No wonder astronomers on earth can through their telescopes plainly see these black shadows contrasting with the glaring brightness on the other side.

A "mass of rocks" I have said; but as, with our powers of rapid movement, we draw near, we find a range of craggy mountains sweeping round in a vast circle. Such a height in Switzerland would demand many hours of hard climbing. But on this small globe attraction is a very different matter from what it is on earth; our weight is so lessened that we can leap the height of a tall house without the smallest difficulty. No chamois ever sprang from peak to peak in his native Switzerland with such amazing lightness as that with which we now ascend these mighty rocks.

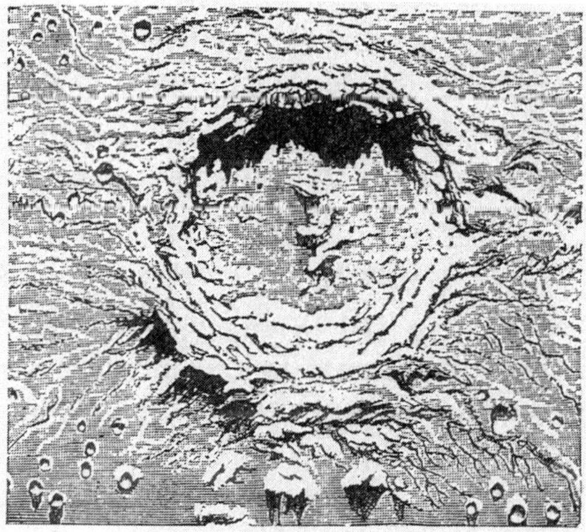

ONE FORM OF LUNAR CRATER.

Ha! what a depth on the other side! We stand looking down into one of the monster craters of the moon. A sheer descent of at least eleven thousand feet would land us at the bottom. Why, Mont Blanc itself is only about fifteen thousand feet in height. And what a crater! Fifty-six miles across in a straight line, from here to the other side, with these lofty rugged battlements circling round, while from the center of the rough plain below a sharp, cone-shaped mountain rises to about a quarter of the height of the surrounding range.

It is a grand sight; peak piled upon peak, crag upon crag, sharp rifts or valleys breaking here and there the line of the narrow, uplifted ledge; all wrapped in silent and desolate calm. There are many such craters as this on the moon, and some much larger.

The sun slowly nears his setting, and sinks behind the opposite range. How we shiver! The last ray of sunlight has gone and already the ground is pouring out its heat into space, unchecked by the presence of air or clouds. The change takes place with marvelous quickness. A deadly chill creeps over all around. A whole fortnight of earth-time must pass before the sun's rays will again touch this spot. Verily the contrasts of climate in the moon, during the twelve long days and nights which make up her year, are startling to human notions.

But though the sun is gone we are not in darkness. The stars shine with dazzling brightness, and the huge body of the earth, always seeming to hang motionless at one fixed point in the sky, gives brilliant light, though at present only half her face is lit up and half is in shadow. Still her shape is plainly to be seen, for she has ever round her a ring of light, caused by the gathered shining of stars as they pass behind her thick atmosphere. She covers a space on the sky more than a dozen times as large as that covered by the full moon in our sky.

It would be worth while to stay here and watch the half-earth grow into magnificent full-earth. But the cold is becoming fearful—too intense for even the

THE EARTH AS SEEN FROM THE MOON.

imagination to endure longer. What must be the state of things on the other side of the moon, where there is no bright earth-light to take the place of the sun's shining, during the long two weeks' night of awful chill and darkness?

It seems probable that a building on the moon would remain for century after century just as it was left by the builders. There need be no glass in the windows, for there is no wind and no rain to keep out. There need not be fireplaces in the rooms, for fuel can not burn without air. Dwellers in a city in the moon would find that no dust can rise, no odors be perceived, no sounds be heard. Man is a creature adapted for life in circumstances which are very narrowly limited. A few degrees of temperature, more or less; a slight variation in the composition of air, the precise suitability of food, make all the difference between health and sickness, between life and death. Looking beyond the moon, into the length and breadth of the universe, we find countless celestial globes, with every conceivable variety of temperature and of constitution. Amid this vast number of worlds with which space teems, are there any inhabited by living beings? To this great question science can make no response save this: We can not tell.

Time for us to wend our way homewards from this desolate hundred-fold arctic scene. We have more to learn by and by about our friend and companion. For the present—enough.

CHAPTER VII.

VISITORS.

We come next to the very largest members of our Solar System.

From time to time in past days—and days not very long past either—people were startled by the sight of a long-tailed star, moving quickly across the sky, called a comet. We see such long-tailed stars still, now and then; but their appearance no longer startles us.

It is hardly surprising, however, that fears were once felt. The great size and brilliancy of some of these comets naturally caused large ideas to be held as to their weight, and the general uncertainty about their movements naturally added to the mysterious notions afloat with respect to their power of doing harm.

A collision between the earth and a comet seemed no unlikely event; and if it happened—what then? Why, then, of course, the earth would be overpowered, crushed, burnt up, destroyed. So convinced were many on this point that the sight of a comet and the dread of the coming "end of the world" were fast bound together in their minds.

Even when astronomers began to understand the paths of some of the comets, and to foretell their return at certain dates, the old fear was not quickly laid to rest. So late as the beginning of the present century, astronomers having told of an approaching comet, other people added the tidings of an approaching collision. "If a collision, then the end of the world," was the cry; and one worthy family, living and keeping a shop in a well-known town on the south coast of England, packed up and fled to America—doubtless under full belief that the destruction of the Old World would not include the destruction of the New.

The nature of these singular bodies is somewhat better known in the present day; yet even now, among all the members of the Solar System, they are perhaps the ones about which we have most to learn. The nucleus, or bright and star-like spot, which, with the surrounding coma or "hair," we sometimes call the "head" of the comet, is the densest and heaviest part of the whole. The comets are of immense size, sometimes actually filling more space than the sun himself, and their tails stream often for millions of miles behind them; nevertheless, they appear to be among the lightest of the members of the Solar System.

This excessive lightness greatly lessens the comet's power of harm-doing. In the rebound from all the old exaggerated fears, men laughed at the notion of so light and delicate a substance working any injury whatever, and even declared that a collision might take place without people on earth being aware of the fact. It is now felt that we really know too little about the nature of the said substance to be able to say what might or might not be the result of a collision. A certain amount of injury to the surface of the earth might possibly take place. But of the "end of the world," as likely to be brought about by any comet in existence, we may safely banish all idea.

The word "comet" means "a hairy body," the name having been given from the hairy appearance of the light around the nucleus. About seventeen hundred different comets have been seen at different times by men—some large, some small; some visible to the naked eye, but most of them only visible through telescopes. These hundreds are, there is no doubt, but a very small number out of the myriads ranging through the heavens.

If you were seated in a little boat in mid-ocean, counting the number of fishes which in one hour passed near enough in the clear water for your sight to reach them, you might fairly conclude, even if you did not know the fact, that for every single fish which you could see, there were tens of thousands which you could not see.

Reasoning thus about the comets, as we watch them from our earth-boat in the ocean of space, we feel little doubt that for each one which we can see, millions pass to and fro beyond reach of our vision. Indeed, so long ago as the days of Kepler, that great astronomer gave it as his belief that the comets in the Solar System, large and small, were as plentiful as the fishes in the sea. And all that modern astronomers can discover only tends to strengthen this view.

Why should the comets be called "visitors?" I call them so simply because many of them *are* visitors. Some, it is true, belong to the Solar System. But even in their case, strong doubts are felt whether they were not once visitors from a distance, caught in the first instance by the attraction of one of the larger planets, and retained thenceforward, for a time at least, by the strong attraction of the sun.

Every comet, like every planet, has his own orbit or pathway in the heavens, though the kind of orbit

PASSAGE OF THE EARTH AND THE MOON THROUGH THE TAIL OF A COMET.

varies with different comets. There are, first, those comets which travel round and round the sun in "closed orbits"—that is, in a ring with joined ends. Only the ring is always oval, not round. There are, secondly, those which travel in an orbit which *may* be closed; but if so, the oval is so long and narrow, and the farther closed end is at so great a distance, that we can not speak certainly. There are, thirdly, those which decidedly are mere visitors. They come from the far-oft star-depths, flash once with their brilliant trains of light through our busy Solar System, causing some little excitement by the way, and go off in another direction, never to return.

Only a small part of the orbits of these comets can be seen from earth; but by careful attention astronomers learn something of the shape of the curve in which they travel. It is in that way possible to calculate, sometimes certainly, and sometimes uncertainly, whether a comet may be expected to return, or whether we have seen him for the first and the last time. By looking at *part* of a curve, the rest of which is hidden from us, we are able to judge whether that part belongs to a circle or an oval, or whether the two ends pass away in different directions, and do not join.

The comets, whether members of our family circle or visitors from a distance, are altogether very perplexing. They are often extremely large, yet they are always extremely light. They reflect the sun's brightness like a planet, yet in some measure they seem to shine by their own light, like a star. They obey the attraction of the sun, yet he appears to have a singular power of driving the comets' tails away from himself.

For, however rapidly the comet may be rushing round the sun, and however long the tail may be, it is almost always found to stream in an opposite direction from the sun. An exception to this rule was seen in the case of a certain comet with two tails, one of which did actually point towards the sun; but the inner tail may have been only a "jet" of unusual length, like in kind to the smaller jets often thus poured out from the nucleus.

Very curious changes take place in comets as they journey, especially as they come near the sun. One was seen in the course of a few days to lose all his hair, and also his tail. Another was seen to break into two pieces, both of which pieces at last disappeared. Sometimes the one tail divides into two tails.

Traveling, as the comets do, from intense cold into burning heat, they are very much affected by the violent change of climate. For the paths of the comets are such long ovals, or ellipses, that, while they approach the sun very closely in one part of their "year," they travel to enormous distances in the other part.

"Halley's Comet," which takes seventy-six of our years to journey round the sun, comes nearer to him than Venus, and goes farther away from him than Neptune. As this comet draws gradually closer, he has to make up for the added pull of the sun's increasing attraction by rushing onward with greater and greater rapidity, till he whirls madly past the sun, and then, with slowly slackening speed, journeys farther and farther away, creeps at length lazily round the farther end of his orbit in the chill, dark, neighborhood of Neptune, and once more travels towards the sun with growing haste.

"Encke's Comet" has a year of only three and a half of our years, so he may be said to live quite in our midst. But many comets travel much farther away than the one named after Halley. It is calculated of some that, if they ever return at all, it can not be for many hundreds of years.

"Newton's Comet," seen about two centuries ago, has a journey to perform of such length that he is not expected again to appear for several thousand years. Yet, at the nearest point in his orbit, he approached the sun so closely, that the heat which he

PASSAGE OF THE COMET OF 1843 CLOSE TO THE SUN (FEBRUARY 27TH, 10 HOURS, 29 MINUTES.)

endured was about two thousand times that of red-hot iron. Changes were seen to be taking place in his shape, as he drew near to the sun, and disappeared. Four days he was hidden in the sun's rays. He vanished, with a tail streaming millions of miles behind him. He made his appearance again with a tail streaming millions of miles in front of him. But how this wonderful movement took place is beyond man's power to explain.

The comet of 1843 was one of the most attractive seen during the present century. It was first observed in March, and it appeared with a suddenness which had quite a startling effect. It was an imposing object in the southern regions. The comet was seen in Italy on the 28th of February; at Washington, on the 6th of March; at Oporto, on the 14th; but owing to unfavorable weather, it was not visible in England, or any of the northern countries of Europe, previous to the 17th. A little after sunset on that day the tail was observed in the western sky, but the head had already sunk below the horizon. The whole of the comet appeared on the following evenings for a short time, for it was traveling away from the sun with great velocity, having doubled the solar orb before it became visible; and about the beginning of April it finally disappeared.

The appearance of this startling stranger, as observed at Washington, is thus described by Lieutenant Maury, of the Hydrographical Office in that city: "On Monday morning, March 6th, our attention was called to a paragraph in the newspapers, stating that a comet was visible near the sun at midday with the naked eye. The sky was clear; but not being able to discover any thing with the unassisted eye, recourse was had to the telescope, but with no better success. About sunset in the evening, the examination was renewed with great diligence, but to no purpose. The last faint streak of day gilded the west; beautiful and delicate fleeces of cloud curtained the bed of the sun; the upper sky was studded with stars, and all hopes of seeing the comet that evening had vanished. Soon after we had retired, the officer of the watch announced its appearance in the west. The phenomenon was sublime and beautiful. The needle was greatly agitated, and a strongly marked pencil of light was streaming up from the path of the sun in an oblique direction to the southward and eastward; its edges were parallel. It was 30° long. Stars could be seen twinkling through it, and no doubt was at first entertained that this was the tail of the comet."

The tail, as seen in northerly countries, spread over an arc of the heavens of about 40°; but in southern latitudes it extended to from 60° to 70°. It had an absolute length of two hundred millions of miles; so that, had it been coiled around the earth like a serpent, it would have girdled it eight thousand times at the equator. This comet approached still nearer the sun than that of Newton, and must therefore have been exposed to a heat of greater intensity. Its center is computed to have been within a hundred thousand miles of the solar surface; and according to Sir John Herschel's calculations, it was then exposed to a heat equal to that which would be received by an equal portion of the earth's service, if it were subject to the influence of forty-seven thousand suns, placed at the common distance of the actual sun. It is difficult to conceive how a flimsy substance in such circumstances could escape being entirely dissipated. But such was its velocity that it wheeled round the sun in less than two hours.

The general question of the probability and the consequences of a collision with a comet may be legitimately entertained. With reference to the first point, it can not be denied that collision is possible; but, at the same time, it is so extremely improbable that it may be safely dismissed from apprehension. The fact of such an event not having been experienced in the known course of terrestrial history is surely some guarantee against its occurrence. Another may be found in the small volume of the earth and of comets when compared with the immensity of space in which they move. According to the well-understood principles of probabilities, Arago has calculated that, upon the appearance of a new comet, the odds are as 281,000,000 to 1, that it will not strike against our globe. But even supposing collision to occur, all that we know of the constitution of comets justifies the conclusion that the encounter would involve no terrestrial convulsion, nor any result incompatible with full security to life and happiness.

So much for the largest members of our circle—largest, though lightest; members some, visitors others. Now we turn to the smallest.

CHAPTER VIII.

LITTLE SERVANTS.

If you walk out any night after dark, and watch the bright stars shining in a clear sky—shining as they have done for ages past—you will probably see, now and then, a bright point of light suddenly appear, dart along a little distance, and as suddenly vanish. That which you have seen was not the beginning of a story, but in ninety-nine cases out of a hundred it was the end of a story. The little shooting-star was in existence long before you saw him, whirling through space with millions of little companions. But he has left them all, and dropped to earth. He is a shooting-star no longer.

If such a journey to the moon as the one described two chapters back were indeed possible, the voyage aloft would hardly be so easily and safely performed as is there taken for granted. Putting aside the thought of other difficulties, such as lack of conveyance and lack of air, there would be the danger of passing through a very considerable storm of missiles—a kind of "celestial cannonade"—which, to say the least, would prove very far from agreeable.

These "starlets" and "meteor planets," as they have been called, are not visible in a normal condition, because of their minuteness. But on entering our atmosphere they are rendered luminous, owing to the heat evolved by the sudden and violent compression of the air in front of the moving body. According to this view, shooting-stars, which simply dart across the heavens, may be regarded as coming within the limits of the atmosphere, and carried out of it again, by their immense velocity, passing on in space. Meteoric

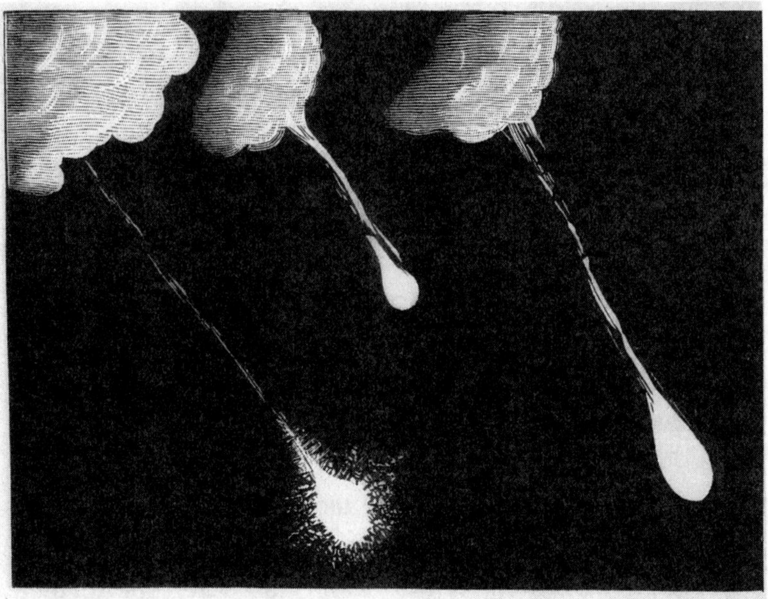

METEOR EMERGING FROM BEHIND A CLOUD (NOV. 23, 1877).

showers may result from an encounter with a group of these bodies, while aërolites are those which come so far within the sphere of the earth's attraction as to fall to its surface. More than two thousand years ago the Greeks venerated a famous stone which fell from the heavens on the river Ægos.

It will scarcely be believed what numbers of these shooting-stars or meteorites constantly fall to the earth. As she travels on her orbit, hurrying along at the rate of nineteen miles each second, she meets them by tens of thousands. They too, like the earth, are journeying round the great center of our family. But they are so tiny, and the earth by comparison is so immense, that her strong attraction overpowers one after another, drags it from its pathway, and draws it to herself.

And then it falls, flashing like a bright star across the sky, and the little meteorite has come to his end. His myriads of companions, hastening still along their heavenly track—for the meteorites seem to travel commonly in vast flocks or companies—might, had they sense, mourn in vain for the lost members of their family.

Any one taking the trouble to watch carefully some portion of the sky after dark, may expect to see each hour about four to eight of these shooting-stars—except in the months of August and November, when the number is much larger. About six in an hour does not sound a great deal. But that merely means that there have been six in one direction, and near enough for you to see. Somebody else, watching, may have seen six in another direction; and somebody else, a few miles away, may have seen six more. It is calculated that, in the course of every twenty-four hours, about four hundred millions of meteorites fall to earth, including those visible only through telescopes.

This is rather startling. What if you or I should some day be struck by one of these solid, hard little bodies, darting as they do towards earth with speed swifter than that of a cannon-ball? True, they are not really stars, neither are they really planets. But they are, to say the least, often much larger than a cannon-ball, and a cannon-ball can destroy life.

Four hundred millions every twenty-four hours! Does it not seem singular that we do not see them constantly dropping to the ground? The truth is, we *should* see them, and feel them too, and dire would be the danger to human life, but for a certain protecting something folded round this earth of ours to ward off the peril. That "something" is the earth's atmosphere. But for the thick, soft, strong, elastic air through which the meteorites have to pass, they would fall with fearful violence, often doing terrible mischief. As it is, we are guarded. The shooting-star, drawn by the earth's attraction, drops into her atmosphere, darting with tremendous speed. In consequence of this speed and the resistance of the air, it catches fire. That is when we first see it. The meteorites are believed to appear at a height of about seventy miles, and to disappear at a height of about fifty miles. So that, in one instant's flash, the shooting-star has traveled some twenty miles toward us. Then the light goes out. The little meteorite is burnt. It falls to earth still; but only as fine dust, sinking harmlessly downward.

The meteorites do not always vanish so quickly. Now and then a larger one—too large to be rapidly burnt—does actually reach the ground. If any man were struck by such a stone he would undoubtedly be killed.

When meteorites thus fall to earth they are usually called *aërolites*. Some are found no bigger than a man's fist, while others much exceed this size. There is one, kept carefully in the British Museum, which weighs three tons and a half; and we hear of another, lying in South America, between seven and eight feet in length. Such a sky-visitant would be very unwelcome in any of our towns.

We must remember that, whatever size an aërolite may be when it reaches the earth, it must have been far greater when journeying round the sun, since a good part of it has been burnt away during its rush downward through the earth's atmosphere.

Meteors, or bolides, or fireballs, are of much the same nature as meteorites; but they are larger, longer to be seen, and slower in movement. Also, it is not uncommon for them to burst with a loud explosion. Early in the present century such a meteor visited Normandy. It exploded with a noise like the roll of musketry, scattering thousands of hot stones over a distance of several miles. A great number of them were collected, still smoking. The largest of these stones weighed no less than twenty pounds. Happily no one seems to have been injured. Other such falls have taken place from time to time. Sometimes bright, slowly-moving meteors have been seen, looking as large as the moon.

A remarkable fireball appeared in England on November 6, 1869. This fireball was extensively seen from different parts of the island, and by combining and comparing these observations, we obtain accurate information as to the height of the object and the velocity with which it traveled. It appears that this meteor commenced to be visible at a point ninety miles above Frome, in Somersetshire, and that it disappeared at a point twenty-seven miles over the sea, near St. Ives, in Cornwall. The whole length of its course was about 170 miles, which was performed in a period of five seconds, thus giving an average velocity of thirty-four miles a second. A remarkable feature in the appearance which this fireball presented was the long, persistent streak of luminous cloud, about fifty miles long and four miles wide, which remained in sight for fully fifty minutes. We have in this example an illustration of the chief features of the phenomena of a shooting-star presented on a very grand scale. It is, however, to be observed that the persistent luminous streak is not a universal, nor, indeed, a very common characteristic of a shooting-star.

If we may liken comets to the *fishes* of the Solar System—and in their number, their speed, their varying sizes, their diverse motions, they may be fairly so likened—we may perhaps speak of the meteorites as the *animalcula* of the Solar System. For, in comparison with the planets, they are, in the matter of size, as the animalcula of our ponds in comparison with human beings. In point of numbers they are countless.

Take a single drop of water from some long-stagnant pond, and place it under a powerful microscope. You will find it to be full of life, teeming with tiny animals, darting briskly to and fro. The drop of water is in itself a world of living creatures, though the naked eye of man could never discover their existence. So with the meteorites. There is good reason to believe that the Solar System fairly teems with them.

We talk of "wide gaps of empty space," between the planets; but how do we know that there is any such thing as empty space to be found throughout all the sun's domain?

Not only are the meteorites themselves countless, a matter easily realized, but the families or systems of meteorites appear to be countless also. They, like the systems of Jupiter and Saturn, are each a family within a family—a part of the Solar System, and yet a complete system by themselves. Each circles round the sun, and each consists of millions of tiny meteorites. When I say "tiny," I mean it of course only by comparison with other heavenly bodies. Many among them may possibly be hundreds of feet and even more in diameter, but the greater proportion appear to be much smaller. It is not impossible that multitudes beyond imagination exist, so small in size that it is impossible we should ever see them, since their dying flash in the upper regions of our atmosphere would be too faint to reach our sight.

The earth, traveling on her narrow orbit round the sun, crosses the track of about one hundred of these systems, or rings. Sometimes she merely touches the edge of a ring, and sometimes she goes into the very thick of a dense shower of meteorites. Twice every year, for instance, on the 10th of August and the 11th of November, the earth passes through such a ring, and very many falling stars may be seen on those nights. Numbers of little meteorites, dragged from their orbits and entangled in the earth's atmosphere, like a fly caught in a spider's web, give their dying flash, and vanish. It used to be supposed that the

August and November meteorites belonged to one single system; but now they are believed to be two entirely distinct systems.

The comet discovered on February 27, 1827, by Biela, and ten days later at Marseilles by Gambart, who recognized that it was the same as that of 1772

THE GREAT SHOWER OF SHOOTING-STARS, NOVEMBER 27, 1872.

and 1805, returned six and a half years later, in 1832. In fact it crossed, as we have seen, the plane of the terrestrial orbit at the respectable distance of fifty millions of miles from the earth; but if there was any danger in this meeting, it was rather for it than for us; for it was certainly strongly disturbed in its course. It returned in 1839, but under conditions too unfavorable to enable it to be observed—in the month of July, in the long days, and too near the sun. It was seen again in 1845, on November

25th, near the place assigned to it by calculation, and its course was duly followed. Everything went on to the general satisfaction, when—unexpected spectacle! —on January 13, 1846, *the comet split into two!* What had passed in its bosom? Why this separation? What was the cause of such a celestial cataclysm? We do not know; but the fact is, that instead of one comet, two were henceforth seen, which continued to move in space like two twin-sisters—two veritable comets, each having its nucleus, its head, its coma, and its tail, slowly separating from each other. On February 10th there was already a hundred and fifty thousand miles of space between the two. They would seem, however, to have parted with regret, and during several days a sort of bridge was seen thrown from one to the other. The cometary couple, departing from the earth, soon disappeared in the infinite night.

They returned within view of the earth in the month of September, 1852. On the 26th of this month the twins reappeared, but much farther apart, separated by an interval of twelve hundred and fifty thousand miles.

But this is not the strangest peculiarity which this curious body presented to the attention of astronomers. The catastrophe which was observed in 1846 was only a presage of the fate which awaited it; for now its existence is merely imagined, the truth being that *this comet is lost.* Since 1852 all attempts to find it again have been unavailing.

To be lost is interesting, especially for a comet. But this, doubtless, was not enough; for it reserved for us a still more complete surprise. Its orbit intersects the terrestrial orbit at a point which the earth passes on November 27th. Well, nothing more was thought about it—it was given up as hopeless, when, on the evening of November 27, 1872, there fell from the sky a veritable *rain of shooting stars.* The expression is not exaggerated. They fell in great flakes. Lines of fire glided almost vertically in swarms and showers—here, with dazzling globes of light; there, with silent explosions, recalling to mind those of rockets; and this rain lasted from seven o'clock in the evening till one o'clock next morning, the maximum being attained about nine o'clock. At the observatory of the Roman College, 13,892 were counted; at Montcalieri, 33,400; in England a single observer counted 10,579, etc. The total number seen was estimated at *a hundred and sixty thousand.* They all came from the same point of the sky, situated near the beautiful star Gamma of Andromeda.

On that evening I happened to be at Rome, in the quarter of the Villa Medicis, and was favored with a balcony looking towards the south. This wonderful rain of stars fell almost before my eyes, so to say, and I shall never cease regretting not having seen it. Convalescent from a fever caught in the Pontine Marshes, I was obliged to go into the house immediately after the setting of the sun, which on that evening appeared from the top of the Coliseum to sleep in a bed of purple and gold. My readers will understand what disappointment I felt next morning when, on going to the observatory, Father Secchi informed me of that event! How had he observed it himself? By the most fortunate chance: A friend of his, seeing the stars fall, went to him to ask an explanation of such a phenomenon. It was then half-past seven. The spectacle had commenced; but it was far from being finished, and the illustrious astronomer was enabled to view the marvelous shower of nearly *fourteen thousand* meteors.

This event made a considerable stir in Rome, and the pope himself did not remain indifferent; for, some days afterwards, having had the honor of being received at the Vatican, the first words that Pius IX addressed to me were these: "Have you seen the shower of Danaë?" I had admired, some days before in Rome, some admirable "Danaës," painted by the great masters of the Italian school in a manner which left nothing to be desired; but I had not had the privilege of finding myself under the cupola of the sky during this new celestial shower, more beautiful even than that of Jupiter.

What was this shower of stars? Evidently—and this is not doubtful—the encounter with the earth of myriads of small particles of matter moving in space along the orbit of Biela's comet. The comet itself, if it still existed, would have passed twelve weeks before. It was not, then, to speak correctly, the comet itself which we encountered, but perhaps a fraction of its decomposed parts, which, since the breaking-up of the comet in 1846, would be dispersed along its orbit behind the head of the comet.*

*No doubt can remain of the identity of this swarm of shooting-stars with the comet of Biela. On November 27, 1885, the same encounter occurred. A magnificent shower of stars was observed all over Europe just at the moment when the earth crossed the comet's orbit.

Once in every thirty-three years we have a grand display of meteorites in November; tens of thousands being visible in one single night. The meteorites in that ring have their "year" of thirty-three earthly years, and once in the course of that long year our earth's orbit carries her deep into their midst. In this single November ring there are myriads upon myriads of meteorites, spreading through millions of miles of space.

Yet this system is but one among many. There is no reason whatever to suppose that the streams of meteorites cluster more thickly about the orbit of the earth, than in other parts of the Solar System. No doubt the rest of the planets come across quite as many. Indeed, the wonderful rings of Saturn are probably formed entirely of meteorites—millions upon millions of them whirling round the planet in a regular orbit-belt, lit up by the rays of the sun. Also it is believed that the meteorite families cluster more and more closely in the near neighborhood of the sun, rushing wildly round him, and falling by millions into the ocean of flame upon his surface. It has even been guessed that they may serve in part as fuel to keep up his mighty furnace heat.

There is a curious cone-shaped light seen sometimes in the west after sunset. It is called the "Zodiacal Light," and men have often been much puzzled to account for it. The shining is soft and dim, only to be seen when the sky is clear, and only to be seen in the neighborhood of the sun. This, too, *may* be caused by reflected light from countless myriads of meteorites gathering thickly round the sun.

CHAPTER IX.

NEIGHBORING FAMILIES.

WE have now to take flight in thought far, far beyond the outskirts of our little Solar System. Yes, our *great* Solar System, with its mighty sun, its planets, its moons, its comets and meteorites, its ceaseless motions, its vast distances,—even all this sinks to littleness beside the wider reaches of space which now have to be pictured to our minds. For our sun, in all his greatness, is only a single star—only one star among other stars—and not by any means one of the largest of the stars.

How many stars are there in the sky? Look overhead some cloudless night, and try to count the brilliant points of light. "Millions," you would most likely give as your idea of their number. Yet you would be wrong; for you do not really perceive so many. The stars visible to man's naked eye have been mapped and numbered. It is found that from two to three thousand are, as a rule, the utmost ever seen at once, even on a favorable night, and with particularly good sight.

But what is actually the full number of the stars? Two or three thousand overhead. Five or six thousand round the whole world. So much visible to man's unaided eyes. Ah, but take a telescope, and see through it the opening fields of stars beyond stars. Take a stronger telescope, and note how, as you pierce deeper into space, fresh stars beyond fresh stars shine faintly in the measureless distance. Take the most powerful telescope ever made, and again it will be the same story.

There has been a chart or map drawn of known stars in the northern hemisphere—including those visible in telescopes down to a certain magnitude—containing over three hundred thousand. But that is only a part of even what man can see. Sir William Herschel calculated roughly that the number of stars within reach of his powerful telescope, round the whole earth, amounted probably to something like twenty millions.

Twenty millions of suns! For that is what it really means. Twenty millions of radiant, burning, heavenly bodies—some the same size as our sun; some larger, perhaps very much larger; some smaller, perhaps very much smaller; but all SUNS. And any number of these suns may have, just like our own, families of planets traveling round them, enjoying their light and their heat.

We talk about stars of the first, second, and other magnitudes. Stars can be seen without a telescope as low down as the sixth magnitude; after that they become invisible to the naked eye. This word "magnitude" is rather misleading. "Magnitude" means size, and whatever the real size of the stars may be, they have to our sight no seeming size at all. So when we speak of different *magnitudes*, we really mean different *brightnesses*. The brightest stars are those of the first magnitude, the next brightest those of the second magnitude, and so on. No doubt many a star of the third or fourth magnitude is really much larger than many a star of the first or second magnitude, only being farther away it shines more dimly, or the higher-magnitude star may in itself possess greater natural brilliancy.

Of first-magnitude stars there are altogether about twenty; of second-magnitude stars about sixty-five; of third-magnitude stars about two hundred; and so the numbers increase till of the sixth-magnitude stars we find more than three thousand. These are all that can be commonly seen with the naked eye, amounting to five or six thousand. With telescopes the numbers rise rapidly to tens of thousands, hundreds of thousands, and even millions.

For a long while it was found quite impossible to measure the distances of the stars. To this day the distances of not over two dozen, among all those tens of thousands, have been discovered. The difficulty of finding out the distance of the sun was as nothing compared with the difficulty of finding out the distances of the stars.

No base-line sufficient for the purpose could for years be obtained. I must explain slightly what is meant by a "base-line." Suppose you were on the brink of a wide river, which you had no means of crossing, though you wished to discover its breadth. Suppose there were on the opposite brink a small tree, standing alone. As you stood, you would see the tree seeming to lie against a certain part of the country beyond. Then, if you moved along your bank some fifty paces, the tree would seem to lie against quite a different part of the country beyond.

Now if you had a long piece of string to lay down along the fifty paces you walked, and if two more pieces of string were tied, one from each *end* of the fifty paces, both meeting at the tree, then the three pieces of string would make one large triangle, and the "fifty paces" would be the "base" of your triangle.

If you could not cross the river, you could not of course tie strings to the tree. But having found your *base-line*, and measured its exact length, and having also found the shape of the two angles at its two ends, by noting the seeming change of the tree's position, it would then be quite easy to find out the distance of the tree. The exact manner in which this calculation is made can hardly be understood without some slight knowledge of a science called trigonometry. The tree's distance being found, the breadth of the river would be known.

This mode of measuring distance was found comparatively easy in the case of the moon. But in the case of the sun there was more difficulty, on account of the sun's greater distance. No base-line of ordinary length would make the sun seem to change his position in the sky in the slightest degree. Nor till the very longest base-line on the earth was tried could the difficulty be overcome. That base-line is no less than eight thousand miles long. One man standing in England looking at the sun, and another man standing in Australia looking at the sun, have such a base-line lying between them, straight through the center of the earth.

In the case of the stars this plan was found useless.

So closely has the sky been mapped out, and so exactly is the place of each star known, that the tiniest change would have been at once noticed. Not a star showed the smallest movement. The eight thousand miles of the earth's diameter was a mere point with regard to them.

A bright idea came up. Here was our earth traveling round the sun, in an orbit so wide that in the middle of summer she is over one hundred and eighty millions of miles away from where she is in the middle of winter. Would not that make a magnificent base-line? Why not observe a star in summer and observe the same star again in winter, and then calculate its distance.

This, too, was done. For a long while in vain! The stars showed no signs of change, beyond those due to causes already known. Astronomers persevered, however, and with close and earnest care and improved instruments, success at last rewarded their efforts. A few—only a few, but still a few—of those distant suns have submitted to the little measuring-line of earth, and their distance has been roughly calculated.

Now, what is their distance? Alpha Centauri, the second star which was attempted with success, is the nearest of all whose distance we know. You have heard how far the sun is from the earth. The distance of Alpha Centauri is *two hundred and seventy-five thousand times as much.* Can you picture to yourself that vast reach of space—a line ninety-three millions of miles long, repeated over and over again two hundred and seventy-five thousand times?

But Alpha Centauri is one of the very nearest.

The brilliant Sirius is at least twice as far away. Others utterly refuse to show the smallest change of position. It is with them, as had been said, much the same as if a man were to look at a church-steeple, twenty miles distant, out of one pane in a window, and then were to look at it out of the next pane. With the utmost attention he would find no change of position in the steeple. And like the base-line of two glass panes to that steeple, so is the base-line formed by our whole yearly journey to thousands of distant stars. We *might* measure how far away they are, only the longest base-line within our reach is too short for our purpose.

The planet Neptune has a wider orbit than ours. But even his orbit, seen from the greater number of the stars, would shrink to a single point. After all, how useless to talk of two hundred and seventy-five thousand times ninety-three millions of miles! What does it mean? We can not grasp the thought.

Let us look at the matter from another view. Do you know how fast light travels—this bright light shining round us all day long? Light, so far as we know, does not exist everywhere. It travels to and fro, from the sun to his planets, from the planets to one another; from the sun to the moon, from the moon to the earth, and from the earth to the moon again.

Light takes time to travel. This sounds singular, but it is true. Light can not pass from place to place in no time. Light, journeying through space, is invisible. Only when it strikes upon something, whether a solid body or water or air, does it become visible to our eyes. The shining all round us in the day-time is caused by the sunlight being reflected, not only from the ground, but from each separate particle of air. If we had no atmosphere, we should see still the bright rays falling on the ground, but the sky above would be black. Yet that black sky would be full of millions of light-rays, journeying hither and thither from sun and stars, invisible except where they alight upon something.

The speed of light is far beyond that of an express train, far beyond that of the swiftest planet. In one tick of the clock, Mercury has rushed onward twenty-nine miles. In one tick of the clock, storm-flames upon the surface of the sun will sweep over two or three hundred miles. But in one tick of the clock a ray of light flashes through one hundred and eighty-six thousand three hundred and thirty-seven miles.

One hundred and eighty-six thousand miles! That is the same as to say that, during one single instant, a ray of light can journey a distance equal to about eight times round and round our whole earth at the Equator.

By using this wonderful light-speed as a measurement, we gain clearer ideas about the distances of the stars. A ray of light takes more than eight minutes to pass from the sun to the earth. Look at your watch, and note the exact time. See the hand moving slowly through the minutes, and imagine one single ray of light, which has left the sun when first you looked, flashing onward and onward through space, one hundred and eighty-six thousand miles each second. Eight minutes and a half are over. The ray falls upon your hand. In those few minutes it has journeyed ninety-three millions of miles.

So much for the sun's distance. How about the stars? Alpha Centauri, a bright star seen in the southern hemisphere, is one of our nearest neighbors. Yet each light-gleam which reaches the eye of man from that star, left Alpha Centauri four years and a third before. During four years and a third, from the moment when first it quitted the surface of the blazing sun, it has flashed ceaselessly onward, one hundred and eighty-six thousand miles each second, dwindling down with its bright companion-rays from a glare of brilliancy to a slender glimmer of light till it reaches the eye of man. Four years and a third sounds much, side by side with the eight minutes' journey from the sun. Sound would take more than three millions of years to cross the same abyss. At the constant velocity of thirty-seven miles an hour, an express-train starting from the sun Alpha Centauri would not reach the earth until after an uninterrupted course of nearly seventy-five millions of years.

This is our *neighbor* star. The second, the nearest after it, is nearly double as far, and is found in quite another region of space, in the constellation of Cygnus, the Swan, always visible in our northern hemisphere. If we wish to understand the relative situation of our sun and the nearest two, let us take a celestial globe, and draw a plane through the center of the globe and through Alpha Centauri and 61 Cygni. We shall thus have before us the relation which exists between our position in infinitude and those of these two suns. The angular distance which separates them on the celestial sphere is 125°. Let us make this drawing, and we shall discover certain rather curious particulars. In the first place, these two nearest stars are in the plane of the Milky Way, so that we can also represent the Milky Way on our drawing; again, this celestial river is divided into two branches, precisely in the positions occupied by these two nearest stars, the division remaining marked along the whole interval which separates them. This drawing shows us, further, that if we wish to trace the curve of the Milky Way with reference to the distance of our two stars, it will be nearer to us in the constellation of the Centaur than in that of the Swan; and, in fact, it is probable that the stars of that region of the sky are nearer than those of the opposite region. Another very curious fact is, that both the nearest stars are double.

But look at Sirius, that beautiful star so familiar to us all. The light which reaches you to-night, left his surface from nine to ten years ago. Look at the Little Bear, with the Pole-star shining at the end of his tail. That ray of soft light quitted the Pole-star some forty years ago. Almost half a century it has been speeding onward and ever onward, with ceaseless rapidity, till its vast journey is so far accomplished that it has reached the earth. Look at Capella, another fair star of the first magnitnde. The light which reaches you from her has taken over thirty years to perform its voyage. Ten, thirty, forty years—at the rate of 186,000 miles per second!

These stars are among the few whose distance can be roughly measured. Others lie at incalculable distances beyond. There are stars whose light must, it is believed, have started hundreds or even thousands of years before the soft, faint ray at length reaches the astronomer's upturned eye through the telescope.

Look at Capella, how gently and steadily she shines! You see Capella, not as she is *now*, but as she *was* thirty years ago. Capella may have ceased to exist meantime. The fires of that mighty sun may have gone out. If so, we shall by and by learn the fact—more than thirty years after it happened. Look at the Pole-star. Is there any Pole-star now? I can not tell you. I only know there *was* one, forty years ago, when that ray of light started on its long journey. Stars have ceased to shine before now. What may have happened in those forty years, who can say? Look at that dim star, shining through a powerful telescope with faint and glimmering light. We are told that in all probability the tiny ray left its home long before the time of Adam.

There is a strange solemnity in the thought. Hundreds of years ago—thousands of years ago—some say, even tens or hundreds of thousands of years ago! It carries us out of the little present into the unknown ages of a past eternity.

If the neighboring stars are placed at tens and hundreds of trillions of miles from us, it is at quadrillions, at quintillions of miles that most of the stars lie which are visible in the sky in telescopic fields. What suns! what splendors! Their light comes from such distances! And it is these distant suns which human pride would like to make revolve round our atom; and it was for our eyes that ancient theology declared these lights, invisible without a telescope, were created! No contemplation expands the thought, elevates the mind, and spreads the wings of the soul like that of the sidereal immensities illuminated by the suns of infinitude. We are already learning that there is in the stellar world a diversity no less great than that which we noticed in the planetary world. As in our own Solar System the globes already studied range from 6 miles in diameter (satellites of Mars) up to 88,000 miles (Jupiter)—that is to say, in the proportion of 1 to 14,000—so in the sidereal system the suns present the most enormous differences of volume and brightness: 61 Cygni, the stars numbered 2,398 in the catalogue of Lalande, and 9,352 in the catalogue of Lacaille, and others of the eighth or ninth magnitude, are incomparably smaller or less luminous than Sirius, Arcturus, Capella, Canopus, Rigel, and the other brilliants of the firmament.

CHAPTER X.

OUR NEIGHBORS' MOVEMENTS.

How high! how distant! how mighty! How little we know about them, yet how overwhelming the little we know, and how wonderful that we do know it!

We have now to consider the movements of these distant neighbors—first, their seeming movements; secondly, their real movements.

I have already spoken about the seeming motions of the stars as a whole, once believed to be real, and now known to be only caused by the motions of our earth. For just as the turning of the earth upon her axis makes the sun seem to rise every morning in the east, and to set every evening in the west, so that same continued turning makes the stars seem to rise every evening in the east and to set every morning in the west.

When we speak of the stars as rising in the east, we do not mean that they all rise at one point in the east, but that all rise, more or less, in an easterly direction—northeast, east, and southeast. So also with respect to the west. It is to the east and west of the earth as a whole that they rise and set—not merely to the east and west of that particular spot on earth where one man may be standing. All night long fresh stars are rising, and others are setting, and if it were not for the veil of light made by the sunshine in our atmosphere, we should see the same going on all day long as well.

There are some constellations, or groups of stars, always visible at night in our northern hemisphere; and there are some constellations never visible to us, but only seen by people living in the southern hemisphere—in Australia, for instance. There are other constellations which appear in summer and disappear in winter, or which appear in winter and disappear in summer. This change is caused by our earth's journey round the sun. It is not that the constellations have altered their place in the heavens with respect to the other constellations, it is merely that the earth has so altered its position in the heavens that the groups of stars which a short time ago were above the horizon with the sun by day are now above the horizon without him by night.

Mention has been a good many times made of the axis of the earth ending in the North and South Poles. If this axis were carried straight onward through space, a long, slender pole passing upwards into the sky without any bend, from the North Pole in one direction and from the South Pole in the other,—this would be the pole of the heavens. The places of the stars in the sky are counted as "so many degrees" from the North and South Celestial Poles, just as the places of towns on earth are counted as "so many degrees" from the North and South Poles of earth. There are atlases of the sky made as well as atlases of the earth.

The constellation of the Great Bear is known to all who have ever used their eyes at all to watch the heavens. Almost equally well known are the two bright stars in this constellation named the Pointers, because, taken together, they point in nearly a straight line to a certain important star in the end of the Little Bear's tail, not very distant.

This star, important less from its brightness than from its position, lies close to that very spot in the heavens where the celestial North Pole passes. It is called the Pole-star. Night after night, through the year, it there remains, all but motionless, never going below the horizon for us in the northern hemisphere, or northern half of the earth; never rising above the

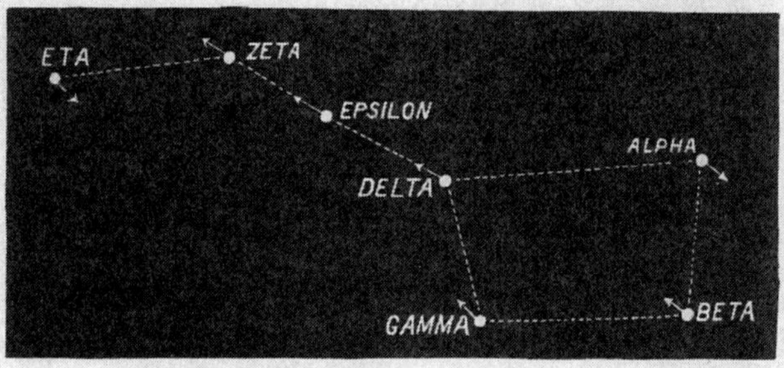

CONSTELLATION OF THE GREAT BEAR.

horizon for those in the southern hemisphere. It shines ever softly and steadily in its fixed position. If you travel further south, the Pole-star sinks downward towards the horizon. If you travel further north, the Pole-star rises higher above the horizon. If you were at the North Pole, you would see the Pole-star exactly overhead.

Very near the Pole-star is the constellation of the Great Bear, with Cassiopeia nearly opposite on the other side of the Little Bear, and other groups between the two, completing the circle. These constellations do not, to us who live in the northern hemisphere, rise or set; for they simply move in a circle round and round the Pole-star, never going below the horizon. All day and all night long this circling movement continues, though only visible at

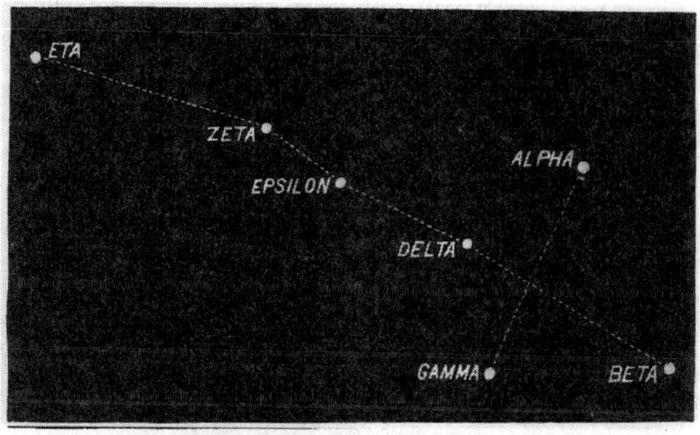

THE GREAT BEAR 50,000 YEARS AGO.

night. It is caused entirely by the earth's own motion on her axis.

Lower down, or rather further off from the Pole-star, comes another ring of constellations. These in

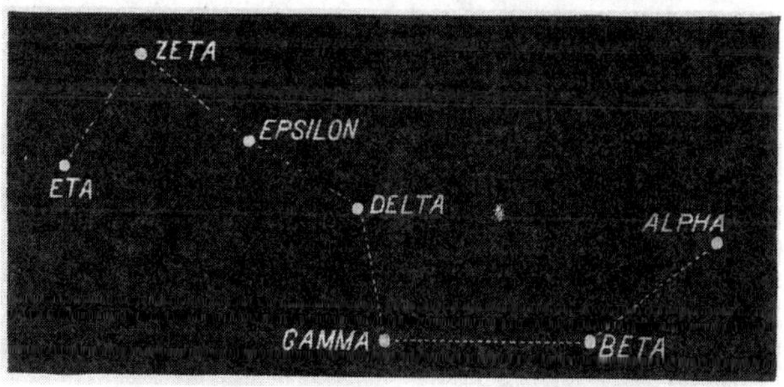

THE GREAT BEAR 50,000 YEARS HENCE.

just the same manner appear to travel round and round the Pole-star. But being further away, each dips in turn below the horizon—or, as we call it, each sets and rises again. And by the time we come to yet another circle of leading constellations, we reach those which are so far affected by the earth's yearly journey as to be only visible through certain months, and to be hidden during other months.

If we could stand exactly at the North Pole, during part of its six months' night, we should see the Pole-star just overhead, and all the constellations circling round it once in every twenty-four hours. Those

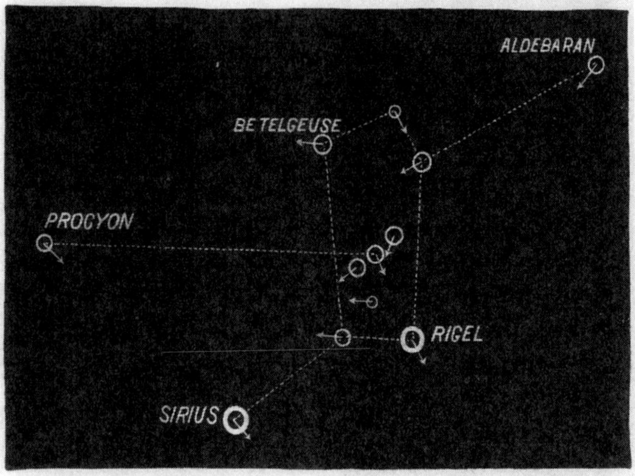

CONSTELLATION OF ORION, AS IT APPEARS NOW.

nearest would move slowly, in a small ring. Those furthest, and lowest down, would in the same length of time sweep round the whole horizon. But the stars would not there seem to rise or set. If we were standing at the South Pole, we should see exactly the same kind of seeming movement, only with altogether a different set of stars. If we were standing on the Equator at night, we should see the rising and setting very plainly. The whole mass of stars would appear to rise regularly and evenly in an easterly direction, to pass steadily across the sky, each taking its own straightforward path, and to set in a westerly direction.

We who are placed midway between the Pole and the Equator, see a mixture of these two motions. Some stars seem to circle round and round, as all would do if we stood at the North Pole. Some stars seem to rise and set, as all would do if we stood at the Equator. So much for the seeming movements of the stars.

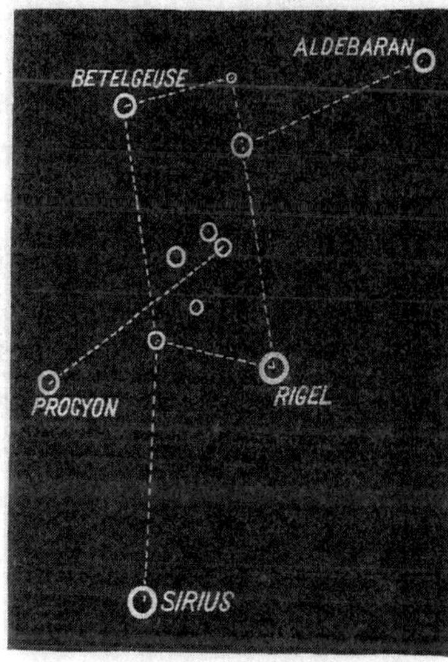

CONSTELLATION OF ORION 50,000 YEARS FROM NOW.

But now, about their real movements. Are the stars fixed, or are they not? These seeming daily and yearly motions do not affect the question, being merely caused by our own motions. Trees and hedges may appear to move as we rush past them in a train, yet they are really fixed.

During a long while, after it was found out that the quick, daily movements of all the stars in company were merely apparent, men believed that they really had no "proper motions"—that is, no movements of their own. For century after century the constellations remain the same. Hundreds of years ago the seven chief stars of the Great Bear shone in company as they shine now. Who could suppose that each one of those seven stars is hurrying on its path through space with a speed exceeding far that of the swiftest express-train? Yet so it is. Hundreds of years ago the grand group of Orion, with belt and sword, gleamed brilliantly night by night as it gleams in these days; and Cassiopeia had her W form, and Hercules and Draco and Andromeda were shaped as they are shaped still. Who would imagine that through those hundreds of years each star of these different constellations was hastening with more or less of speed along its heavenly road? Yet so it is.

Cases have been decisively ascertained of stars changing their places among the other stars by a slow and gradual motion. Three of the most conspicuous of them—Sirius, Arcturus, and Aldebaran—have been proved, by the comparison of modern with some ancient observations, to have experienced a change of place to the southward, to the extent of more than the breadth of the moon in all the three. And during the period of accurate modern measurement, other instances have been ascertained of steady change of place by the effect of proper motion.

But if the stars are thus rapidly moving in all directions, how is it that we do not *see* them move? How is it that, night after night, year after year, century after century, even thousand years after thousand years, the shapes of the constellations remain unaltered?

Suppose you and I were standing on the seashore together, watching the movements of scores of sea-craft, little boats and large boats, steamers, yachts, and ships. Suppose we stood through a full quarter of an hour looking on. Some might move, it is true, very slowly; yet their movements in every case would plainly be seen. There could be no possibility of mistaking the fact, or of supposing them to be "fixed." Just so we see the nearer planets move. Little danger of our supposing them to be "fixed stars."

In the matter of the stars themselves, we must carry our illustration further. Come with me up to the top of that lofty hill on the border of the sea, and let us look from the cliff. We see still the movements among boats and smacks, yachts and steamers, only the increased distance makes the movements seem slower. But our view is widened. Look on the far horizon and see three distant dots, which we know to be ships—one and two close together, and a third a little way off, making a small constellation of vessels. Watch them steadily for a quarter of an hour. You will detect no movement, no increased distance or nearness between any two of the three. The group remains unchanged.

Are they really moving? Of course they are, more or less rapidly, probably with differing speed and in different directions. But at so great a distance, one quarter of an hour is not long enough for their motions to become visible to the naked eye. If we could watch longer—say, for two or three hours—ah, that would make all the difference! If only we could watch longer! But the hundreds, and even thousands of years during which men have watched the stars, sink, at our vast distance, into no more than one quarter of an hour spent in watching the far-off ships from the high hilltop. The motions can not be detected. In ten thousand years you might see something. In fifty thousand years you might see much. But four or five thousand years are not sufficient.

One other mode there is, by means of which the movements of the ships on the horizon might be made plain. Suppose you had no more than the quarter of an hour to spare, but suppose you had at your command a powerful telescope. Then you may practically bring the ships nearer, and by magnifying the small, slow, distant motions, you may make them, as it were, larger, quicker, more easy to see.

Telescopes will do this for us, likewise, in the matter of the stars. By means of telescopes, with the assistance of careful watching and of close calculation it has been found that the stars are really moving quickly, each one in his own pathway. The very speed of some of them has been measured.

Arcturus is one of those stars, the motions of which are most plainly to be seen. In the course of about one thousand years he changes visibly his place in the sky by a space equal to the apparent diameter of the moon. The seeming movement of Arcturus in one thousand years is the same as the seeming width of the round moon that we see.

But the actual speed with which Arcturus rushes through space is said to be no less than fifty-four miles each second, or not far from two hundred thousand miles an hour. That is nearly three times as fast as our own earth's motion round the sun. How enormous the distance must be which can shrink such speed to such seeming slowness!

Another beautiful star, Capella, is believed to travel at the rate of thirty miles each second. The speed of Sirius is slower, being only fourteen miles a second. The Pole-star creeps along at the rate of only one mile and a half each second.

So also with the rest of the stars. There seems good reason to believe that every star we see shining in the heavens, every star visible in powerful telescopes, is perpetually hastening onward. But hastening whither? God knows! We do not.

In all probability not one of the tens of millions of stars which may be seen through telescopes is in repose. This is a matter of conjecture, of reasoning from analogy, and of reasoning also from the working of known laws. We *know*, as a consequence of direct observation, apart from the new spectroscopic method, that at least hundreds are upon the wing. We *assume*, as a matter of the greatest possible likelihood, that all the millions besides, which can not be actually seen to stir, are equally on the move. Knowing what we do know of the laws by which the universe of stars is governed, it seems to us an absolute impossibility that any single star, amid the whole vast host, can be or could be permanently at rest.

If by any means a star were brought to repose—what then would happen? It would inevitably start off again, drawn by the attraction of other stars. From whatever direction the strongest pull came, the impulse would be given. Lengthened repose would be out of the question. And this, it seems to us, must be true, not of one star only, here or there, but of every star in the enormous host of radiant suns which make up the mighty Stellar System.

Our forefathers, one thousand years ago, could not

measure the precise positions of individual stars, as astronomers now are able to do. They had no modern observatories, no telescopes, no spectroscopes, no photographic appliances. These methods of observation we shall explain in another chapter. Their measurements at best were rough, their scientific knowledge was crude. Had we any such accurate observations handed down from one thousand years ago as are made in these days, we should no doubt see clearly many slight differences in the positions of many stars which are not now apparent. But even then we should see no changes sufficient in amount to affect the general outlines of the leading constellations.

A star, which in the course of a century makes visible advance over a space in the sky equal to only a small portion of the breadth of the full moon, is looked upon as a fast voyager. One hundred times this degree of movement, if it took place in a considerable number of stars in our sky, would not in centuries very materially change the face of our midnight heavens.

And all motions which would in the remotest degree affect the shapes of constellations, must be sideway motions. Those line-of-sight motions, of which the spectroscope alone tells us, could never have been discovered by simple observation of the sky. Until the new method came to light we had no means whatever of perceiving such movements among the stars.

Suppose you are looking at two men in the distance, upon a wide, flat plain. One of the two is walking very slowly *across* your line of vision. The second man is moving very slowly straight *towards* you. If you watch with care you may find out both the movements. The sideway walking will be apparent first and most easily, because, as the man moves, he has constantly a fresh part of the horizon behind him. But in time the advance of the second man towards you will also become apparent; for although he is seen still against precisely the same spot on the horizon, he slowly occupies a larger spot on the retina of your eye,—in other words, he seems to grow bigger. And that, as you know from long experience, can only mean increasing nearness.

If a star seemed to grow larger as it drew nearer, we should then be able to perceive that movement also. But no star in the sky ever does seem to grow any larger. Every star is to us but one point of light. It may be rushing towards us at an enormous rate of speed, yet still as a single point it remains, always at the same point in our sky. Therefore we have no chance of perceiving its movement; or rather, we *had* no chance until the discovery of this new method. In fact, the very nearest known star is at so enormous a distance that, supposing it to be coming towards us at the rate of one hundred miles each second, it would still gain, in the course of a century, only one-fortieth part more of brightness than it has now. It would not increase at all in apparent size.

When we leave behind us the thought of starry motions, as we faintly detect them at this great distance, and picture to ourselves the actual far off whirl of all those glorious suns, the effect upon the mind is overwhelming. Stars are found to be rushing hither and thither, at every degree of speed, in every imaginable direction: stars to right, and stars to left; stars towards us, and stars away from us; stars alone, and stars in company,—all this, and more, deciphered out of the tiny gleam of quivering light, which streams through the vast abyss of space from each distant orb to earth. So real star-motions were known—first, through telescopic observation; secondly, through spectroscopic observation; and now, lastly, photography has stepped in, bringing with it much increase of exactitude.

But if all the stars are moving, what of our sun? Our sun is a star. And our sun also is moving. He is pressing onward, in a wide sweep through space, bearing along with him his whole vast family—planets, satellites, comets, meteorites—round or towards some far distant center. For aught we know, every star in the heavens may have a like family traveling with him.

In infinite space the stars are strewn in immense clusters, like archipelagoes of islands in the ocean of the heavens. To go from one star to another in the same archipelago light takes years; to pass from one archipelago to another it takes thousands of years. Each of these stars is a sun similar to ours, surrounded, doubtless, at least for the most part, by worlds gravitating in its light; each of these planets possesses, sooner or later, a natural history adapted to its constitution, and serves for many ages as the abode of a multitude of living beings of different species. Attempt to count the number of stars which people the universe, the number of living beings who are born and die in all these worlds, the pleasures and pains, the smiles and tears, the virtues and vices! Imagination, stop thy flight!

The sun is not one of the most quickly-moving stars. His rate of speed has not been found out with any certainty, but it is believed to be about four or five miles a second.

And where are we going? This has been in part discovered. If you and I were driving through a forest of trees, we should see the trees on each side of us seeming to move backward, while behind they would close together, and in front they would open out.

Astronomers—and first among them, William Herschel—reasoned that if our Solar System were really in motion, we ought to be able to see these changes among the stars. And some such changes have become visible through careful watching—not so much those ahead and behind as those at the sides.

It is not actually so simple a matter as looking at the trees in a forest, because the trees would be at rest, whereas each star has his own particular real motion, as well as his seeming change of place caused by our sun's motion. It is more like moving in a small steamer at sea, among hundreds of other craft, each of which is going on its own way, at the same time that all on either side seem to move backward because we are moving forward.

So each movement had to be noted, and the real motions had to be separated from the seeming backward drift of stars to the right and left of the sun's pathway. The result of all this is that the sun, with his planets, is found to be hastening towards a certain far-off constellation named Hercules.

It is a strange and unexpected fact, but absolutely true, that each sun of space is carried along with a velocity so rapid that a cannon-ball represents rest in comparison; it is at neither a hundred, nor three hundred, nor five hundred yards per second that the earth, the sun, Sirius, Vega, Arcturus, and all the systems of infinitude travel: it is at ten, twenty, thirty, a hundred thousand yards a second; all run, fly, fall, roll, rush through the void—and still, seen as a whole, all seems in repose.

The immense distance which isolates us from all the stars reduces them to the state of motionless lights apparently fixed on the vault of the firmament. All human eyes, since humanity freed its wings from the animal chrysalis, all minds since minds have been, have contemplated these distant stars lost in the ethereal depths; our ancestors of Central Asia, the Chaldeans of Babylon, the Egyptians of the Pyramids, the Argonauts of the Golden Fleece, the Hebrews sung by Job, the Greeks sung by Homer, the Romans sung by Virgil,—all these earthly eyes, for so long dull and closed, have been fixed, from age to age, on these eyes of the sky, always open, animated, and living. Terrestrial generations, nations and their glories, thrones and altars have vanished: the sky of Homer is always there. Is it astonishing that the heavens were contemplated, loved, venerated, questioned, and admired, even before anything was known of their true beauties and their unfathomable grandeur?

Better than the spectacle of the sea, calm or agitated, grander than the spectacle of mountains adorned with forests or crowned with perpetual snow, the spectacle of the sky attracts us, envelops us, speaks to us of the Infinite. "I have ascended into the heavens, which receive most of His light, and I have seen things which he who descends from on high knows not, neither can repeat," wrote Dante in the first canto of his poem on "Paradise." Let us, like him, rise towards the celestial heights, no longer on the trembling wings of faith, but on the stronger wings of science. What the stars would teach us is incomparably more beautiful, more marvelous, and more splendid than anything we can dream of.

How do such contemplations enlarge and transfigure the vulgar idea which is generally entertained of the world! Should not the knowledge of these truths form the first basis of all instruction which aims at being serious? Is it not strange to see the immense majority of human beings living and dying without suspecting these grandeurs, without thinking of learning something of the magnificent reality which surrounds them?

Where the sun and his planets will journey in future ages no living man can say. Indeed, though it is a question which does not lack interest to a thoughtful mind, yet there are numberless other questions about centuries near at hand which concern man far more nearly. The history of the Universe, and the history of this Earth of ours, must have advanced many broad stages before our sun and his attendant planets can have traveled so far that any change will be apparent in the shape of the star-constellations which spangle our sky.

CHAPTER XI.

MORE ABOUT THE SOLAR SYSTEM.

WE have now reached a point where it ought not to be difficult for us to picture to ourselves, with something of vividness, the general outlines of the Solar System. Awhile ago this Solar System was a very simple matter in the eyes of astronomers. There was the great sun fixed in the center, with seven planets circling round him—seven of course, it was said, since seven was the perfect number—and a few moons keeping pace with some of the planets, and an occasional comet, and a vast amount of black, empty space. But astronomers now begin to understand better the wonderful richness of the system as a whole; the immense variety of the bodies contained in it; the perpetual rush and stir and whirl of life in every part. Certainly there is no such thing as dull stagnation throughout the family.

First, we have the great, blazing, central sun; not a sun at rest, as regards the stars, but practically at rest as regards his own system, of which he is always head and center. Then come the four smaller planets, rapidly whirling round him, all journeying in the same direction, and all having their oval pathways lying on nearly the same flat plane in space. Then the broad belt of busy little planetoids. Then the four giant planets,—Jupiter nearly five times as far as our earth from the sun; Saturn nearly twice as far as Jupiter; Uranus nearly twice as far as Saturn; Neptune as far from Uranus as Uranus from Saturn,—all keeping on very nearly the same level as the four inner planets.

And between and about these principal members of

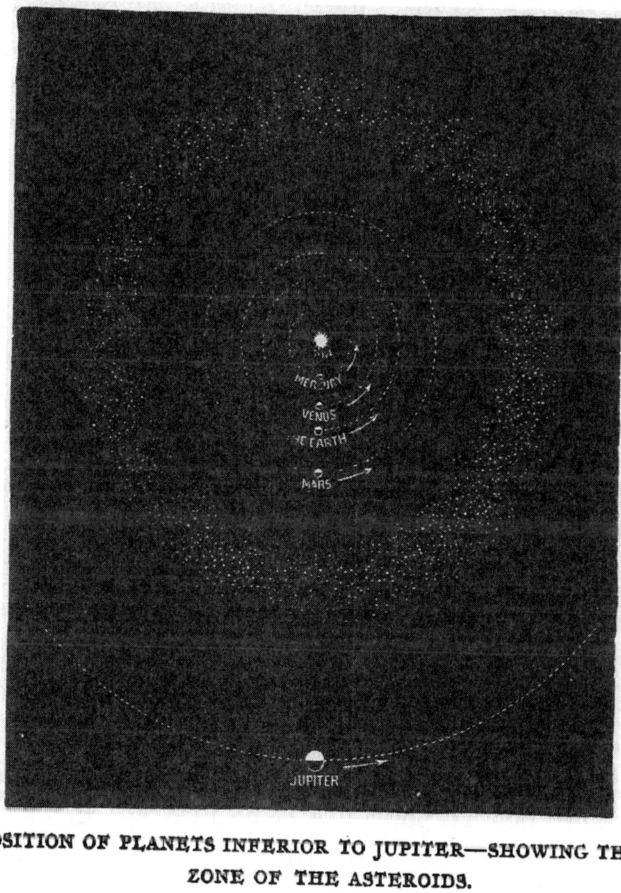

POSITION OF PLANETS INFERIOR TO JUPITER—SHOWING THE ZONE OF THE ASTEROIDS.

the system, with their accompanying moons, we have thousands of comets flashing hither and thither, with long, radiant trains; and myriads of meteorites, gathered often into dense, vast herds or families, but also scattered thickly throughout every part of the system, each tiny ball reflecting the sun's rays with its little glimmer of light.

Broad reaches of black and empty space! Where are they? Perhaps nowhere. We are very apt, in our ignorance, to imagine that where we see nothing, there must of necessity be nothing. But, for aught we know, the whole Solar System, not to speak of sky-depths lying beyond, may be bright with reflecting bodies great and small, from the mighty Jupiter down to the fine diamond-dust of countless meteorites. In this earth of ours we find no emptiness. Closer and closer examination with the microscope only shows tinier and yet tinier wonders of form and life, each perfect in finish. Not of form only, but of *life*. How about that matter as regards the Solar System? Is our little world the one only spot in God's great universe which teems with life? Are all other worlds mere barren, empty wastes? Surely not. We may safely conclude that life of one kind or another has been, is, or will be, upon our brother and sister worlds.

The same reasoning may be used for the distant stars—those millions of suns lying beyond reach of man's unassisted eyes. Are they formed in vain? Do their beams pour uselessly into space, carrying light, warmth, and life-giving power to nothing? Surely around many of them, as around our sun, must journey worlds; and not only worlds, but worlds containing life.

We shall have to speak of this matter again as we go on. Whether men and women like ourselves could live on the other planets is another question. In some cases it looks doubtful, in some it would seem to be impossible. But the endless variety of life on this earth—life on land, life in the air, life in water, life underground, life in tropical heat, life in arctic cold—forbids us in anywise to be positive as to what may or may not be.

If no animals that we know could exist there, animals that we do not know might be found instead. Any one of the planets may, and very likely does, abound with life. Nay, the very meteorites themselves, before they catch fire and burn in our air, *may* be the homes of tiny animalcula, altogether different from anything we see on earth. We can not fancy life without air, but neither could we fancy life in water, if we had not seen and known it to be possible.

I have spoken of the probable *brightness* of the Solar System as a whole. We are so apt to think of things merely as we see them with our short sight, that it is well sometimes to try to realize them as they actually are. Picture to yourself the great central sun pouring out in every direction his burning rays of light. A goodly abundance of them fall on our earth, yet the whole amount of light and heat received over the whole surface of this world is only the two thousand-millionth part of the enormous amount which he lavishly pours into space. How much of that whole is wasted? None, though God gives his gifts with a kingly profusion which knows no bounds. Each ray has its own work to do. Millions of rays are needed for the lighting and nourishing and warming of our companion-planets, while others are caught up by passing comets, and myriads flash upon swift, tiny meteorites. Of the rays not so used, many pass onwards into the vast depths beyond our system, and dwindle down into dim, starlike shining till they reach the far-off brother-stars of our sun.

Have they work to do there? We can not tell. We do not know how far the sun's influence reaches. As head and center, he reigns only in his own system. As a star among stars, a peer among his equals, he may, for aught we can tell, have other work to do.

In an early chapter, mention was made of the earth's three motions, two only being explained. First, she spins ceaselessly upon her axis. So does the sun, and so do the planets. Secondly, she travels ceaselessly round and round the sun in her fixed orbit. So does each one of the planets. Thirdly, she journeys ceaselessly onward through space with the sun. So also do the rest of the planets. These last two movements, thought of together, make the earth's pathway rather perplexing at first sight. We talk of her orbit being an ellipse or oval; but how can it be an ellipse, if she is always advancing in one direction?

The truth is, the earth's orbit is and is not an ellipse. As regards her yearly journey round the sun, roughly speaking, we may call it an ellipse. As regards her movement in space, it certainly is not an ellipse.

Think of the Solar System, with the orbits of all the planets, as lying *nearly flat*—in the manner that hoops might be laid upon a table, one within another. The asteroids, comets, and meteorites do not keep to the same level; but their light weight makes the matter of small importance.

Having imagined the sun thus in the center of a large table—a small ball, with several tiny balls traveling round him on the table at different distances—suppose the sun to rise slowly upwards, not directly up, but in a sharp slant, the whole body of planets continuing to travel round, and at the same time rising steadily with him.

By carefully considering this double movement, you will see that the real motion of the earth—as also of each of the planets—is not a going round on a flat surface to the same point from which she started, but is a corkscrew-like winding round and round upwards through space. Yet as regards the central sun, the shape of the orbit comes very near being an ellipse, if calculated simply by the earth's distance from him at each point in turn of her pathway through the year.

An illustration may help to explain this. On the deck of a moving vessel, you see a little boy walking steadily round and round the mast. Now is that child moving in a circle, or is he not? Yes, he is. No, he is not. He walks in a circle as regards the position of the mast, which remains always the center of his pathway. But his movement *in space* is never a circle, since he constantly advances, and does not once return to his starting-point. You see how the two facts are possible side by side. Being carried forward by the ship, with no effort of his own, the forward motion does not interfere with the circling motion. Each is performed independently of the other.

It is the same with the earth and the planets.

The sun, by force of his mighty attraction, bears them along wherever he goes—no exertion on their part, so to speak, being needed. That motion does not in the least interfere with their steady circling round the sun.

Just as—to use another illustration—the earth, turning on her axis, bears through space a man standing on the Equator at the rate of one thousand miles an hour. But this uniform movement, unfelt by himself, does not prevent his walking backwards, or forwards, or in circles, as much as he will.

So, also, a bird in the air is unconsciously borne along with the atmosphere, yet his freedom to wheel in circles for any length of time is untouched.

A few words about the orbits of the planets. I have more than once remarked that these pathways are, in shape, not circles, but ellipses. A circle is a line drawn in the shape of a ring, every part of which is at exactly the same distance from the center-point or focus. But an ellipse, instead of being, like a circle, perfectly round, is oval in shape; and instead of having only one focus, it has two foci, neither being exactly in the center. Foci is the plural word for focus. If an ellipse is only slightly oval—or slightly *elliptical* —the two foci are near together. The more oval or *eccentric* the ellipse, the farther apart are the two foci.

You may draw a circle in this manner. Lay a sheet of white paper on a board, and fix a nail through the paper into the board. Then pass a loop of thread —say an inch or an inch and a half in length—round the nail, and also round a pencil, which you hold Trace a line with the pencil, keeping the loop tight, so that the distance of your line from the nail will be always equal, and when it joins you have a circle. The nail in the center is the focus of the circle.

To draw an ellipse, you must fix *two* nails. Let them be about half an inch apart; pass a loop over both of them, and again placing a pencil point within the loop, again trace a line carefully all round, keeping the thread drawn tight. This time an oval instead of a circle will appear. By putting the nails nearer together or farther apart, you may vary as you will the shape of the ellipse.

In the orbits of the earth and the planets, all of which are ellipses in shape, the sun is not placed in the exact center, but in one of the two foci, the second being empty. So at one time of the year the planet is nearer to the sun than at another time. Our earth is no less than three millions of miles nearer in winter than she is in summer—speaking of the winter and summer of the northern hemisphere. Three millions of miles is so tiny a piece out of ninety-three millions of miles, that it makes little or no difference in our feelings of heat or cold.

The orbits of the comets are ellipses also, but ellipses often so enormously lengthened out, that the two foci are almost—if one may so speak—at the two *ends* of the oval. To draw a good comet orbit, you must fix the two nails on your paper some five or six inches apart, with a loop of thread just large enough to slip over them both, and to allow the pencil to pass round them. When your ellipse is drawn, you must picture the sun in the place of one of the two nails, and you will see how, in their pathways, the comets at one time pass very near the sun, and at another time travel very far away from him.

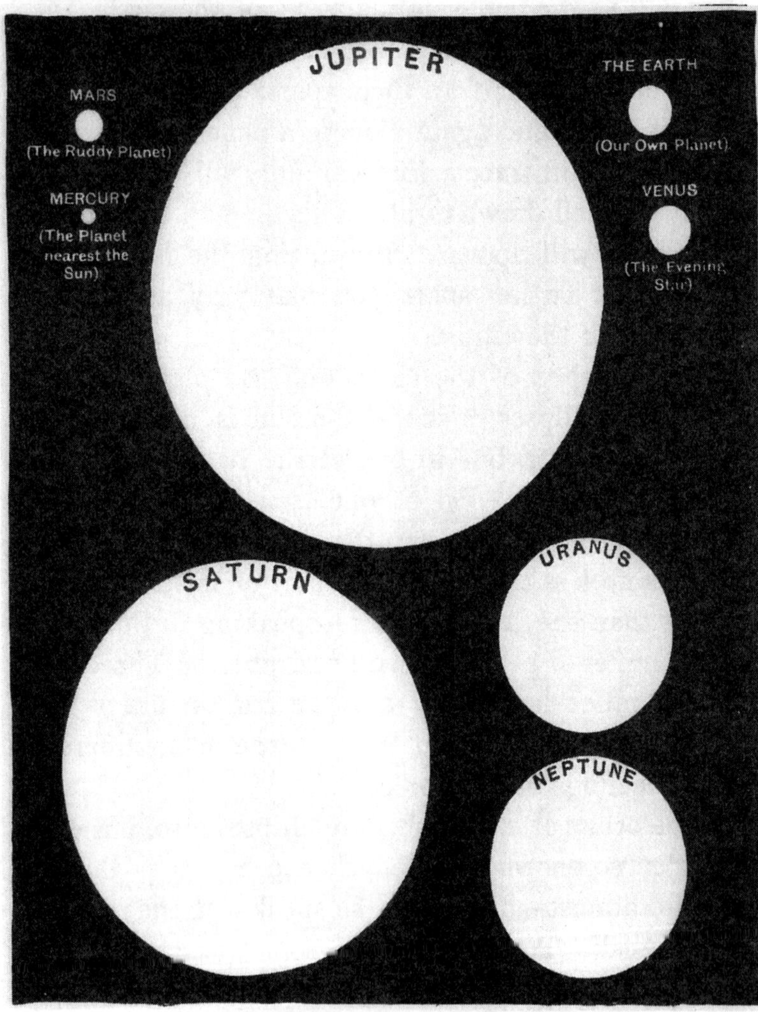

COMPARATIVE SIZE OF THE PLANETARY WORLDS.

It is generally found in families, not only that the parent or head of the family has great influence over all the members, but that each member has influence over each other member. Brother influences brother, and sister influences sister.

This, too, we find in the Solar System. Not only does the sun, by his powerful attraction, bind the whole family together, but each member of the family attracts each other member. True, the force of the sun's attraction is overpowering in amount compared with others. The sun attracts the planets, and the planets attract the sun; but their feeble pulling is quite lost in the display of his tremendous strength.

Among themselves we see the power more plainly. The earth attracts the moon, keeping her in constant close attendance; and the moon attracts the earth, causing a slight movement on her part, and also causing the tides of the sea. Each planet has more or less power to hinder or help forward his nearest brother-planet. For instance, when Jupiter on his orbit draws near the slower Saturn on his orbit, Saturn's attraction pulls him on, and makes him move faster than usual; but as soon as he gets ahead of Saturn then the same attraction pulls him back, and makes him go more slowly than usual. Jupiter has the same influence over Saturn; and so also have Saturn and Uranus over one another, or Uranus and Neptune.

In early days astronomers were often greatly puzzled by these quickened and slackened movements, which could not be explained. Now the "perturbations" of the planets, as they are called, are understood and allowed for in all calculations. Indeed, it is by means of this very attraction that Neptune was discovered, and the planets have actually been weighed. What a wonderful difference we find in this picture of the Solar System, as we now know it to be, from the old-world notion of our earth as the center of the universe!

When we think of all the planets, and of the magnificent sun; when we pass onward in imagination through space, and find our sun himself merely one twinkling star amid the myriads of twinkling stars scattered broadcast through the heavens, while planets and comets have sunk to nothing in the far distance,—then indeed we begin to realize the unutterable might of God's power! Why, our earth and all that it contains may be regarded as but one grain of dust in the wide universe.

CHAPTER XII.

MORE ABOUT THE SUN.

NOT among the least of the wondrous things of creation are the tremendous disturbances taking place upon the surface of the sun—that raging, roaring sea of flame.

A good many explanations have been from time to time offered as to the dark spots seen to move across the face of the sun. Some one or more of these explanations may be true; but we know little about the matter. A sun-spot does not commonly consist of merely one black patch. There is the dark center, called the *umbra*—plural, *umbræ*. There is the grayish part surrounding the umbra, called the *penumbra*. Also, in the center of the umbra there is sometimes observable an intensely black spot, called the *nucleus*. Sometimes a spot is made up of nucleus and umbra alone, without any penumbra. Sometimes it is made of penumbra alone, without any umbra. Sometimes in one spot there are several umbræ, with the gray penumbra round the whole, and gray bridges dividing the umbræ.

The enormous size of these spots has been already described in an earlier chapter. Fifty thousand to one hundred thousand miles across is nothing unusual. In the year 1839 a spot was seen which measured no less than one hundred and eighty-six thousand miles in diameter.

One explanation proposed was, that the sun might be a cool body, covered over with different envelopes or dense layers of cloudy form, one above another. The inside envelope—or, as some say, the inside atmosphere—would then be thick and dull-colored, protecting the solid globe within and reflecting light, but having none of its own. The next envelope would be one mass of raging, burning gases—the *photosphere*, in fact. The outer envelope would be a transparent, surrounding atmosphere, lighted up by the sea of fire within. A sun-spot would then consist of the tearing open of one or more of these envelopes, so as to give glimpses of the gray, inner atmosphere, or even of the dark, cool globe at the center.

There may be some truth in this explanation; but the notion of a cool and dark body within is now pretty well given up. The apparent blackness of a spot-nucleus does not prove actual blackness or absence of heat. A piece of white-hot iron, held up against the sun, looks black; and it may be merely the contrast of the glowing photosphere which makes the nucleus seem so dark. It is even believed that the blackest parts may be the most intensely hot of all.

Another proposed explanation was of dark clouds floating in the sun's atmosphere. There, again, are found difficulties, particularly in the fact that, as the spots move across the sun, the changes which regularly take place in their appearance make it pretty clear that their shape is not flat, but hollow and cave-like. The changes here spoken of are seeming changes of shape, caused by change of position.

There are also real changes constantly taking place. Although the spots often keep their general outlines long enough to be watched across the face of the sun, and even to be known again after spending nearly a

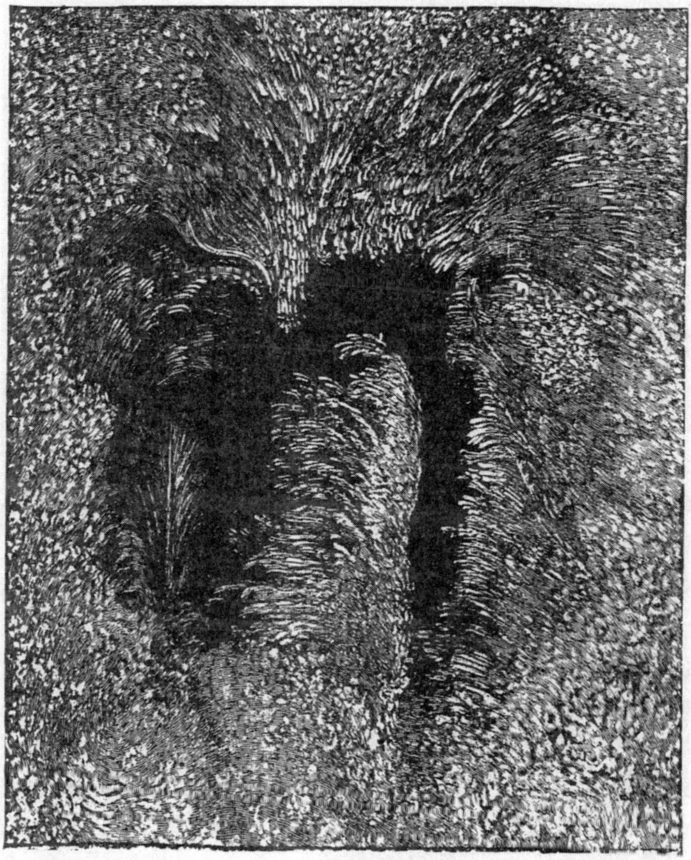

A TYPICAL SUN-SPOT.

fortnight hidden on the other side, still they are far from being fixed in form.

The alterations are at times not only very great, but very rapid. Sometimes in a single hour of watching, an astronomer can see marked movement going on—as you or I might in an hour observe movements slowly taking place in a high layer of clouds. For movement to show at all in one hour, at so immense a distance, proves that the actual rate of motion must be very great.

The first great fact which was got from the study of these spots was this, that this great sun is very much like our own earth, in so far as it rotates on an axis in exactly the same way that our earth does. Not only do the spots change their position on the face of the sun, in consequence of the sun's rotation on its axis, but they change very much from day to day, and even from hour to hour; so that we have evidence not only that the sun is rotating like our earth, but that the atmosphere of the sun is subjected to most tremendous storms—storms so tremendous, in fact, that the fiercest cyclones on our own earth are not for one moment to be compared with them. The fact that the spots do really move with the sun, and are really indentations, saucer-like hollows, in the photosphere, is shown by the appearance, which is always presented by a spot when it is near the edge of the sun. You know if you take a dinner-plate, and look it full in the face, it is round; but if you look at it edgeways, it is not round. Take two different views of the same spot: In one, you look the sun-spot straight in the face, and you see into it and can learn all about it; but in the other, when it has nearly gone round the corner, and is disappearing on the sun's edge, you see it in the same way that you would see a plate looked at edgeways.

These sun-cyclones must indeed be of terrific force

and extent, compared with anything we see on earth. It was calculated that the speed of movement perceived in one spot was about three hundred and sixty-three miles each second.

And still all this is nothing, or almost nothing, in comparison with the real power of the sun! The liquid state of the ocean; the gaseous state of the atmosphere; the currents of the sea; the raising of the clouds, the rains, storms, streams, rivers; the calorific value of all the forests of the globe and all the coal-mines of the earth; the motion of all living beings; the heat of all humanity; the stored-up power in all human muscles, in all the manufactories, in all the guns,—all that is almost nothing compared with that of which the sun is capable. Do we think that we have measured the solar power by enumerating the effects which it produces on the earth? Error! profound, tremendous, foolish error! This would be to believe still that this star has been created on purpose to illuminate terrestrial humanity. In reality, what an infinitesimal fraction of the sun's total radiation the earth receives and utilizes! In order to appreciate it, let us consider the distance of ninety-three millions of miles which separates us from the central star, and at this distance let us see what effect our little globe produces, what heat it intercepts. Let us imagine an immense sphere *traced at this distance from the sun*, and entirely surrounding it. Well, on this gigantic sphere, the spot intercepted by our little earth is only equivalent to the fraction $\frac{1}{2,188,000,000}$; that is to say, that the dazzling solar hearth radiates all round it through immensity a quantity of light and heat two thousand one hundred and thirty-eight million times more than that which we receive, and of which we have just now estimated the stupendous effects. The earth only stops in its passage the *two thousand millionth part of the total radiation.*

Sometimes the storms or outbursts come in the shape of a bright spot instead of a dark one.

A SOLAR ERUPTION.

Two astronomers were one day watching the sun from two different observatories, when they saw such an event take place. An intense and dazzling spot of light burst out upon the surface of the sun—so intense, so dazzling, as to stand quite apart from the radiant photosphere. To one astronomer it looked like a single spot, while the other saw two spots close together. In about a minute the light grew more dim, and in five minutes all was over. But in those five minutes the spot or spots had traveled a distance of thirty-five thousand miles.

It was a very remarkable thing that the *magnets* on earth—those delicate little needles which point so steadily yet perseveringly towards the North Pole—seemed to be strongly agitated by the distant solar outburst. This brings us to another interesting fact.

The spots on the sun are not always the same in number. Sometimes they are many, sometimes they are few. Long and close watching has made it clear that they pass through a regular *order* of changes: some years of many spots being followed by other years of less and less spots; then some years of very few spots being followed by other years of more and more spots,—decrease and increase being seemingly regular and alternate.

This turn or *cycle* of changes—from more to less, and then from less to more again—is found to run its course about once in every eleven years, with some variations.

Now, it has long been known that the magnetic-needle goes through curious variations. Though we speak of it as pointing always north, yet it does not always so point exactly. Every day the needle is found to make certain tiny, delicate motions, as if faintly struggling to follow the daily movements of the sun—just a little towards the east, or just a little towards the west. These tiny motions, having been long watched and measured, were found to go through a regular course of changes—some years more, and some years less, waxing and waning by turns. It was discovered that the course of changes from more to less, and from less to more again, took place in about eleven years.

These two things, you see, were quite independent of one another. Those who watched the sun-spots were not thinking of the magnets, and those who watched the magnets were not thinking of the sun-spots. But somebody did at last happen to think of both together. He was laughed at, yet he took the trouble carefully to compare the two. And, strange to say, he found that these two periods in the main agreed—the eleven years of alternate changes in the number of sun-spots, and the eleven years of alternate changes in the movements of the magnetic-needle. When the spots are most, the needle moves most. When the spots are least, the needle moves least. So much we know. But to explain the why and the wherefore is beyond our power.

This study of the magnetism of our wandering planet is very interesting, and one which is still very little known. Here is a weak needle, a slip of magnetic iron, which, with its restless and agitated finger, incessantly seeks a region near the north. Carry this needle in a balloon up to the higher aerial regions, where human life begins to be extinguished; shut it up in a tomb closely separated from the light of day; take it down into the pit of a mine, to more than a thousand yards in depth,—and incessantly, day and night, without fatigue and without rest, it watches, trembles, throbs, seeks the point which attracts it across the sky, through the earth, and through the night. Now—and here is a coincidence truly filled with notes of interrogation—the years when the oscillation of this innocent little steel wire is strongest are the years when there are more spots, more eruptions, more tempests in the sun;

and the years when its daily fluctuations are weakest are those when we see in the day-star neither spots, eruptions, nor storms. Does there exist, then, a magnetic bond between the immense solar globe and our wandering abode? Is the sun magnetic? But the magnetic currents disappear at the temperature of red-hot iron, and the incandescent focus of light is at a temperature incomparably higher still. Is it an electrical influx which is transmitted from the sun to the earth across a space of ninety-three millions of miles? On September 1, 1859, two astronomers—Carrington and Hodgson—were observing the sun, independently of each other; the first on a screen which received the image; the second directly through a telescope,—when, in a moment, a dazzling flash blazed out in the midst of a group of spots. This light sparkled for five minutes above the spots without modifying their form, as if it were completely independent, and yet it must have been the effect of a terrible conflagration occurring in the solar atmosphere. Each observer ascertained the fact separately, and was for an instant dazzled. Now, here is a surprising coincidence: at the very moment when the sun appeared inflamed in this region the magnetic instruments of the Kew Observatory, near London, where they were observing, manifested a strange agitation; the magnetic needle jumped for more than an hour as if infatuated. Moreover, a part of the world was on that day and the following one enveloped in the fires of an aurora borealis, in Europe as well as in America. It was seen almost everywhere,—at Rome, at Calcutta, in Cuba, in Australia, and in South America. Violent magnetic perturbations were manifested, and at several points the telegraph-lines ceased to act. Why should these two curious events not be associated with each other? A similar coincidence was observed on August 3, 1872, by Professor Charles A. Young, of Princeton, N. J.: a paroxysm in the solar chromosphere, magnetic disturbances everywhere.

There is a very singular appearance seen upon the sun which must not be passed over without mention. Some astronomers speak of the whole surface as being *mottled* all over with a curious rough look when examined through a powerful telescope. This "mottling" is described by various observers in various ways. One speaks of "luminous spots shaped like rice-grains." Another, of "luminous spots resembling strokes made with a camel's-hair pencil." Another, of "luminous objects or granules." Others, of "multitudes of leaves," "nodules," "crystalline shapes," "leaves or scales, crossing one another in all directions, like what are called spills in the game of spillikens." They have also been pictured as "certain luminous objects of an exceedingly definite shape and general uniformity of size, whose form is that of the oblong leaves of a willow-tree." These cover the whole disk of the sun, excepting the space occupied by the spots, in countless millions, and lie crossing each other in every imaginable direction.

In size they are said to be about one thousand miles long, by two or three hundred broad, but they vary a good deal. Where there is a spot, the willow-leaves at its edge are said to point pretty regularly towards the center. Whether they are, as they seem to be, solid in form; whether they are, as some suppose, the chief source of the sun's light and heat; whether they lie on his surface or float in his atmosphere; what is their real nature, and what is their real use,—about these questions we are at present quite in the dark.

We have next to think a little more about the edge or limb of the sun, and the stormy flames and outburst there seen. Until quite lately the only time for observing such appearances was during a total eclipse of the sun. Lately, by means of the new instrument called the "spectroscope," it has been found possible to take observations when no eclipse is going on.

SUN-FLAMES, MAY 3, 1892.

A few words of explanation as to eclipses of the sun seem needful, before going further. An eclipse of the sun is caused simply by the round body of the moon passing exactly between the sun and the earth, so as to hide the sun from us.

Let there be a candle on the table, while you stand near. The rays of light from the candle fall upon your face. Now move slowly, to and fro, a round ball between you and the candle. So long as it is not precisely in the line between—so long as it is a little higher, or a little lower, or a little to one side—then you can still see the flame. Once let the ball come just between the light and your eyes, and you see it no more. In other words the candle-flame is eclipsed—hidden, veiled, cut off—by the ball.

It may seem curious at first sight that the moon, which is very small compared with the sun, should have power to cover the sun. But remember the difference of the distance. The sun is very far, and the moon is very near. Any small object very near will easily hide from your sight a large object at a considerable distance. You may hold up a shilling-piece at arm's length, and make it cover from sight a man, or even a house, if the latter be far enough away. The sun at a distance of ninety-three millions of miles, and the moon at a distance of two hundred and forty thousand miles, have to our vision the same seeming size. So, when the moon glides between, her round face just about covers the sun's round face.

If the moon were traveling exactly in the same plane as the plane of the earth's orbit, an eclipse would be a very common affair indeed. But the plane of the moon's orbit being not quite the same as the plane of the earth's orbit, she passes sometimes a little above, and sometimes a little below, the exact spot where she would hide the sun's rays from us. Now and then, at certain intervals, she goes just between. And so well is the moon's path in the heavens understood that astronomers can tell us, long years beforehand, the day and the hour an eclipse will take place.

An eclipse of the sun is sometimes partial, sometimes total, sometimes annular. In a partial eclipse, the moon does indeed pass between, but only so as to hide from us *part* of the sun. She is a little too low, or a little too high, to cover his face. In a total eclipse, the moon covers the sun completely, so that for a few minutes the bright photosphere seems blotted out from the heavens, a black round body, surrounded by light, taking its place. In an annular eclipse, the moon in like manner crosses the sun, but does not succeed in covering him entirely, a rim of bright photosphere showing round the black moon. For in an annular eclipse, the moon, being a little farther away from the earth than at the time of a total eclipse, has too small a disk quite to hide the sun's disk.

The blackness of the moon during an eclipse is caused by the fact that her bright side is turned towards the sun, and her dark side toward us. An eclipse of the sun can take place only at new moon, never at full moon. At her full, the moon is outside the earth's orbit, away from the sun, and can not by any possibility pass between.

Eclipses, like comets, have always been interpreted as the indication of inevitable calamities. Human vanity sees the finger of God making signs to us on the least pretext, as if we were the end and aim of universal creation. Let us mention, for example, what passed even in France, with reference to the announcement of an eclipse of the sun, on August 21, 1560. For one, it presaged a great overthrow of States and the ruin of Rome; for another, it implied another universal deluge; for a third, nothing less would result than a conflagration of the globe; finally, for the less excited, it would infect the air. The belief in these terrible effects was so general that, by the express order of the doctors, a multitude of frightened people shut themselves up in very close cellars, well heated and perfumed, in order to shelter themselves from these evil influences. Petit relates that the decisive moment approached, that the consternation was at its height, and that a parish priest of the country, being no longer able to confess his parishioners, who believed their last hour had come, was obliged to tell them in a sermon "not to be so much hurried, seeing that, on account of the wealth of the penitents, the eclipse had been postponed for a fortnight." These good parishioners found no more difficulty in believing in the postponement of the eclipse than they had in believing in its unlucky influence.

History relates a crowd of memorable acts on which eclipses have had the greatest influence. Alexander, before the battle of Arbela, expected to see his army routed by the appearance of a phenomenon of this kind. The death of the Athenian general Nicias and the ruin of his army in Sicily, with which the decline of the Athenians commenced, had for their cause an eclipse of the moon. We know how Christopher Columbus, with his little army, threatened with death by famine at Jamaica, found means of procuring provisions from the natives by depriving them in the evening of the light of the moon. The eclipse had scarcely commenced when they supplied him with food. This was the eclipse of March 1, 1504. We need not relate other facts of this nature, in which history abounds, and which are known to every one.

Eclipses no longer cause terror to any one, since we know that they are a natural and inevitable consequence of the combined motions of the three great celestial bodies—the sun, the earth, and the moon; especially since we know that these motions are regular and permanent, and that we can predict, by means of calculation, the eclipses which will be produced in the future as well as recognize those which have occurred in the past.

The following description of the total eclipse of 1860 will be found interesting. Mr. Lowe, who observed it at Santander, Spain, thus writes:

"Before totality commenced, the colors in the sky and on the hills were magnificent beyond all description. The clear sky in the north assumed a deep indigo color, while in the west the horizon was first black like night. In the east the clear sky was very pale blue, with orange and red, like sunrise. On the shadow creeping across, the deep blue in the north changed, like magic, to pale, sunrise tints of orange and red, while the sunrise appearance in the east had changed to indigo. The darkness was great; the countenances of men were of a livid pink. The Spaniards lay down, and their children screamed with fear; fowls hastened to roost, ducks clustered together, pigeons dashed against the houses; flowers closed; many butterflies flew as if drunk, and at last disappeared. The air became very humid, so much so that the grass felt to one of the observers as if recently rained upon."

CHAPTER XIII.

YET MORE ABOUT THE SUN.

HAVING seen something of storms taking place on the sun's photosphere, we must next give our attention to storms taking place at his edge. But it should be remembered that the said edge, far from being a mere rim to a flat surface, is a kind of horizon-line—is, in fact, just that part of the photosphere which is passing out of or coming under our sight. The surface there is, in kind, the same as the surface of the broad disk facing us. In watching outbursts at the edge of the sun, we have a side view instead of a bird's-eye view.

In the year 1871, Professor Charles A. Young, of Princeton, was looking at a large hydrogen cloud on the edge of the sun. When I speak of a "cloud," it must not be supposed that anything like a damp, foggy, earthly cloud is meant. This solar cloud was a huge mass of red-hot gas, about one hundred thousand miles long, rising to a height of fifty thousand miles from the sun's surface, and appearing to rest on glowing pillars of fire.

The professor, while watching, was called away for half an hour. He came back, expecting to find things much as he had left them. Instead of this, a startling change had taken place. The whole mass of glowing fire seemed to have been actually "blown to shreds" by some tremendous outburst from below.

In place of the motionless cloud were masses of scattered fire, each from about four thousand to fourteen thousand miles long, and a thousand miles wide.

As the professor gazed, these "bits" of broken cloud rose rapidly upwards, away from the surface of the sun. When I say "rapidly," I mean that the real movement, which the professor could calculate, was rapid. The seeming movements were of course slow, and over a small space. The actual motions were not tardy, for in ten minutes these huge, fiery cloud-pieces rushed upwards to a height of two hundred thousand miles from the edge of the sun, moving at a rate of at least one hundred and sixty-seven miles each second. Gradually they faded away.

But what caused this sudden change? Just before the professor was interrupted, he had noticed a curious little brilliant lump—a sort of suspicious thunder-cloud appearance—below the quiet, bright cloud. And after this tremendous shattering, the little bright lump rose upwards into a huge mass of rolling flame, reaching like a pyramid to a height of fifty thousand miles. In the course of a few minutes these enormous flames could be seen to move and bend, and to curl over their gigantic tips. But they did not last long. At half-past twelve the professor had been called away; by half-past two the rolling flames completely vanished.

Now, whatever may be the full explanation of this sight, there is no doubt that on that day was observed from earth a tremendous outburst, compared with which our mightiest volcanoes are like the sputtering of a farthing dip beside a roaring furnace. The awful force and greatness of such a solar eruption are more than we can possibly picture to ourselves. At our distance we may catch a faint glimpse of what is going on, and calculate speed of movement. But vividly to realize the actual terrific grandeur of what took place is past our power.

Possibly this was much the same kind of outburst as that seen by the two English astronomers; only theirs was a bird's-eye view, as it were, looking down on the top of the sight, while the professor had a side-view, certainly much the best for observation.

It does not follow from what he saw that the eruption must have taken place exactly at the "edge" of the sun. Probably it happened near the edge. All he could say, was that the flames rose fifty thousand miles, and the pieces of cloud were carried two hundred thousand miles, away from the edge. The eruption may have begun on the other side of the sun, at any distance from the horizon-edge where it first became visible to earthly eyes.

Also, while the professor found that the shattered cloudlets moved at a rate of about one hundred and sixty-seven miles each second, it is calculated that the first fearful outburst must have caused movement, near the surface of the sun, at a rate of at least three hundred miles each second. Probably the hydrogen cloud was borne upwards along with a vast mass of fragments flung out from the sun. We are here upon doubtful ground; but this tremendous power of eruption in the sun, and of driving matter out of and away from his surface, should not be forgotten.

Though such a sun-storm as that just described is not often to be seen, yet there are at all times certain strange red prominences, or glowing flames, rising up here and there from the sun's "limb." Doubtless they rise also from other parts of the photosphere, though they are only visible to us when near enough to the edge to stand out beyond it.

Seen during an eclipse, these prominences have clear, sharp outlines, and are usually bright rose-red in color. They are described as sometimes wide and low, sometimes tall and slender; sometimes jagged, sometimes regular; sometimes keeping long the same shape, sometimes changing quickly in a few minutes. They are said to be like flames, like mountains, like the teeth of a saw, like icebergs, like floating cloudlets.

As to their height, from fifty to eighty thousand miles is nothing unusual. We must not speak of Mont Blanc or Mount Everest here. Jupiter placed bodily on the surface of the sun, beside such a fire-mountain, would not far overtop it. The earth, Venus, Mars, and Mercury, would lie like little toy-balls at its foot. And these are common-sized sun-flames. One was measured which reached to the enormous height of two hundred thousand miles. The spectroscope shows these solar prominences or jets to be made—at least in part—of burning hydrogen gas.

Beyond the sierra or chromatosphere—that border of rippling, crimson flame-billows round the edge of the sun, with red flame-mountains rising out of it here and there—beyond these, stretches the corona. The corona, as seen from earth, is a bright, far-reaching glory of light, shining round the sun in a total eclipse. The moon then comes between he sun and the earth, her dark, round body creeping over the face of the sun till the bright photosphere is completely covered. But the sierra and the tall, red flames stand out from behind the black moon, and the beautiful, soft corona-light stretches far beyond.

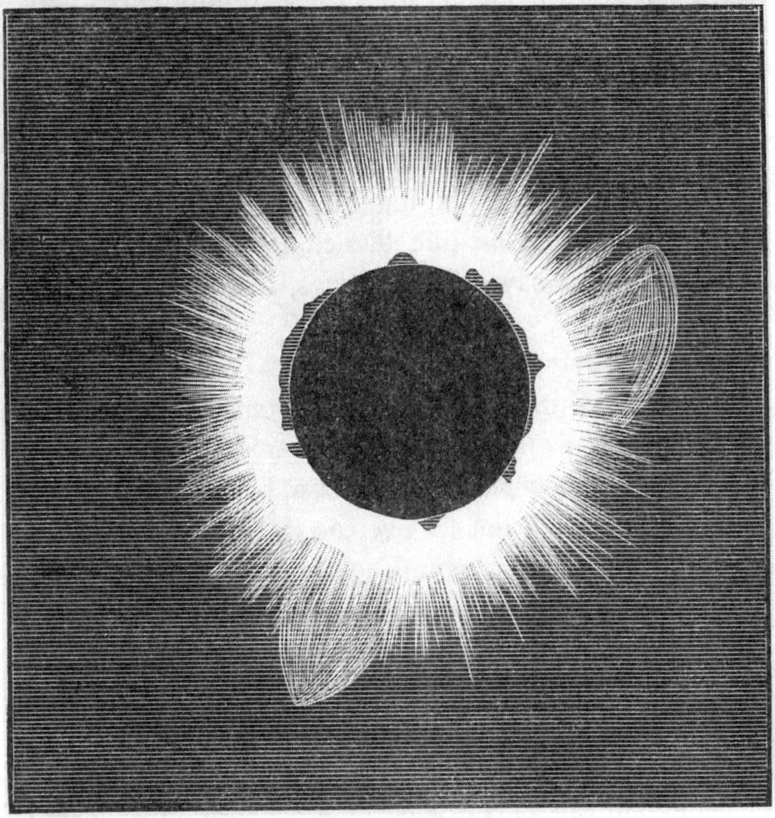

SOLAR CORONA AND PROMINENCE.

It was long doubted whether the corona really belonged to the sun or to the moon. There seems now no doubt that it is a part of the sun.

Various descriptions of the corona have been given at different times, as observed during different eclipses. It has been seen as a steady, beamy, white cloud behind the moon, showing no flickering. It has been seen marked with bright lines of light, and seeming to move rapidly round and round. It has been seen silvery white, sending off long streams of brightness. It has been seen in the form of white light, with bluish rays running over it. It has been seen with entangled jets of light, like "a hank of thread in disorder." It has been seen silvery-white again, with a faint tinge of greenish-violet about the outer edge. It has been seen from a high mountain-top as a mass of soft, bright light, "through which shot out, as if from the circumference of the moon, straight, massive, silvery rays, seeming distinct and separate from each other, to a distance of two or three diameters of the lunar disk, the whole spectacle showing as upon a background of diffused, rose-colored light."

Majestic, indeed, are the proportions of some of those mighty flames which leap from the surface of the sun, yet these flames flicker, as do our terrestrial flames, when we allow them time comparable to their gigantic dimensions. Drawings of the same prominence often show great changes in a few hours, or even less. The magnitude of the changes could not be less than many thousands of miles, and the actual velocity with which such masses move is often not less than one hundred miles a second. Still more violent are the solar convulsions, which some observers have been so fortunate as to behold, when from the sun's surface, as from a mighty furnace, vast incandescent masses are projected upwards. All indications point to the surface of the sun as the seat of the most frightful storms and tempests, in which the winds sweep along incandescent vapors.

The corona consists of two parts—the inner and brighter corona, the outer and fainter corona. The shape of the whole seems to change much at different times. The outer edge is usually blurred and indistinct, fading gently away.

Many explanations have been suggested. At one time the corona was supposed to be a solar atmosphere, reflecting light like our own atmosphere. Some have thought the light might be caused by countless myriads of meteorite systems, revolving in the close neighborhood of the sun. Some suppose it may be owing, in part at least, to solar eruptions, and the pouring outward of burning gas and matter. But our knowledge of the true nature of the corona is yet in its infancy.

A few closing words as to the size and weight of the sun. In diameter, eight hundred and fifty-eight thousand miles, and in bulk equal to one million two hundred and eighty thousand earths, his weight is in proportion less. Our earth is about four times as dense as the sun. If her size were increased to the sun's size, her density being the same as now, she would be very much heavier than the sun, and would attract much more strongly. Still, though the sun is of lighter materials than the earth, his immense size gives him weight equal to seven hundred and fifty times as much as all the planets put together.

The attraction on the surface of the sun is also very great—so great that we can hardly picture it to ourselves. If life exists there at all—supposing it possible that any kind of life can be in such a fiery atmosphere—it must be life very different from any known in this world. A man who on earth weighs one hundred and sixty pounds, and walks lightly erect, would, on the sun, lie helplessly bound to the ground, crushed by his own overpowering weight. It is said that a cannon-ball, reposing on the sun, if lifted one

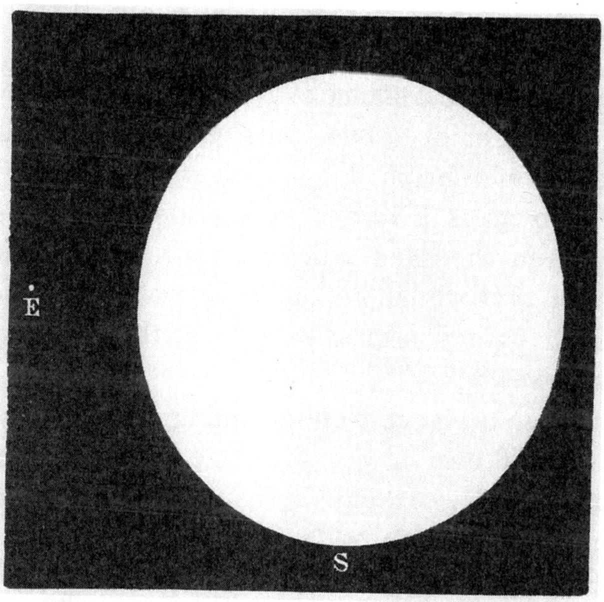

COMPARATIVE SIZE OF THE EARTH AND SUN.

inch and allowed to fall, would dash against the ground with a speed three times greater than that of our fastest express-trains. For weight on earth is merely caused by the amount of force with which the earth draws downward a body towards herself—a force greater or less according to the density of that body. So weight on the sun would be immensely increased by his immensely greater power of attraction.

It is an interesting question how far the sun's attractive influence reaches effectually through space. The nearer a body is to the sun, the greater the attraction which he exercises over it. At the distance of the planet Mercury, a speed of twenty-nine miles each second is needful to overcome or balance it sufficiently for the planet to remain in his orbit. At the distance of the planet Neptune, about three miles each second is enough. If a planet were journeying at four times the distance of Neptune, the speed would need to be not over two miles each second, lest the planet should break loose and wander away. But even two miles a second is no mean speed—more than seven thousand miles an hour. If we come to speak of that which we on earth call rapid motion, we shall gain a clearer idea as to the extent of the sun's power.

Suppose a planet were traveling through space at the rate of one of our fastest express-trains—sixty miles an hour. It has been calculated that, unless the sun's attraction were interfered with and overpowered by some nearer sun, the said planet, though placed at a distance ten or twelve times as great as that of the far-off star Alpha Centauri, would still be forced by the sun's attraction to journey round him in a closed orbit. At such a speed it would not be free to wander off into the depths of space.

CHAPTER XIV.

MORE ABOUT THE MOON.

FROM a globe all fire, all energy, all action, we come to a globe silent, voiceless, changeless, lifeless. So, at least, the moon seems to us. But it will not do to speak too confidently.

True, we can find no trace of an atmosphere in the moon. If there is any atmosphere at all, it must be so thin as to be less than that which we on earth count as actually none. We pump away the air from a glass inclosure in an air-pump, and say the glass is empty. Only it is not quite empty. There is always just a very little air remaining, though so little that fire would not burn and animals could not live in it. Some believe that air, up to that amount, may be found in the moon. But this is much the same as to say there is none at all. For, of course, with either no air or so very little air, life can not possibly exist on the moon. We can not imagine such a thing for a moment. That is just how the matter stands. We "can not imagine," and therefore we conclude it to be an impossibility. As if we knew a hundredth part of the possibilities in any one corner of God's great universe! As if our being unable to picture a thing proves that thing not to exist!

Suppose we had always lived in tropical heat, and had never seen, known, or heard of such a fact as life in Arctic snows. Should we consider it a thing possible? Suppose we had always lived on dry land, with never a sight of sea or river or pond, and never a proof that animal life could exist under water—aye, and that some living animals may be suffocated by air, just as other living animals are suffocated by water. Should we not, in our wisdom, reason out such a state of affairs to be utterly impossible?

There *may* be no life on the moon. It *may* be that she is now passing through a dead, cold, blasted stage, either at the close of some past history, or in preparation for some future history—or both. But, on the other hand, it *may* be that the moon is no less full of life than the earth; only the life must be different in kind, must be something which we do not know any thing at all about.

The moon is very much smaller than the earth. Her diameter is about two-sevenths of the earth's diameter; her entire surface is about two twenty-sevenths of the earth's surface; her size is about two ninety-ninths of the earth's size; and her whole weight is about one-eightieth of the earth's weight. Attraction or gravitation on the surface of the moon is very different from what it is on the earth. Her much smaller bulk greatly lessens her power of attraction. While a man from earth would, on the surface of the sun—supposing he could exist there at all—lie helpless, motionless, and crushed by his own weight, he would on the moon find himself astonishingly light and active. A leap over a tall house would be nothing to him.

The moon, unlike the sun, has no light or heat of her own to give out. She shines merely by reflected light. Rays of sunlight falling upon her, rebound thence, and find their way earthward. This giving of

THE MOON—AN EXPIRED PLANET.

reflected light is not a matter all on one side. We yield to the moon a great deal more than she yields to us. Full earth, seen from the moon, covers a space thirteen times as large as full moon seen from earth.

Perhaps you may have noticed, soon after new moon, when a delicate crescent of silver light shows in the sky, that within the said crescent seems to lie the body of a round, dark moon, only not perfectly dark. It shows a faint glimmer. That glimmer is called earth-shine. The bright crescent shines with reflected sunlight. The dim portion shines with reflected earth-light. What a journey those rays have had! First, leaving the sun, flashing through ninety-three millions of miles to earth, rebounding from earth and flashing over two hundred and forty thousand miles to the dark shaded part of the moon, then once more rebounding and coming back, much wasted and enfeebled, across the same two hundred and forty thousand miles, to shine dimly in your eyes and mine. The popular description of this particular view of the moon is "the old moon in the arms of the new."

Now about the *phases* of the moon; that is, her changes from "new" to "full," and back again to "new." If the moon were a starlike body, shining by her own light, she would always appear to be round. But as she shines by reflected sunlight, and as part of her bright side is often turned away from us, the size and shape of the bright part seem to vary. For, of course, only that half of the moon which is turned directly towards the sun is bright. The other half turned away is dark, and can give out no light at all, unless it has a little earth-shine to reflect.

As the moon travels round the earth, she changes gradually from new to full moon, and then back to new again. "New moon" is when the moon, in her orbit, comes between the sun and the earth. The half of her upon which the sun shines is turned away from us, and only her dark side is towards us. So at new moon she is quite invisible. It is at new moon that an eclipse of the sun takes place, when the moon's orbit carries her in a line precisely between sun and earth.

Passing onwards round the earth, the moon, as we get a little glimpse of her shining side, first shows a slender sickle of light, which widens more and more till she reaches her first quarter. She is then neither between earth and sun, nor outside the earth away from the sun, but just at one side of us, passing over the earth's own orbit. Still, as before, half her body is lighted up by the sun. By this time *half* the bright part and *half* the dark part are turned towards us; so that, seeing the bright quarter, we name it the "first quarter."

On and on round us moves the moon, showing more light at every step. Now she passes quite outside the earth's orbit, away from the sun. Not the slightest chance here of an eclipse of the sun, though an eclipse of the moon herself is quite possible. But more of that presently. As she reaches a point in a line with earth and sun—only generally a little higher or lower than the plane of the earth's orbit—her round, bright face, shining in the sun's rays, is turned exactly towards us. Then we have "full moon."

Still she goes on. Once more her light narrows and wanes, as part of her bright half turns away. Again at the "last quarter," as at the first, she occupies a "sideways" position, turning towards us half her bright side and half her dark side. Then she journeys on, with lessening rim of light, till it vanishes, and once more we have the dark, invisible "new moon."

It was these phases and aspects of the moon which formerly gave birth to the custom of measuring time by months, and by weeks of seven days, on account of the return of the moon's phases in a month, and because the moon appears about every seven days, so to say, under a new form. Such was the first measure of time; there was not in the sky any signal of which the differences, the alternations, and the epochs were more remarkable. Families met together at a time fixed by some lunar phase.

The new moons served to regulate assemblies, sacrifices, and public functions. The ancients counted the moon from the day they first perceived it. In order to discover it easily, they assembled at evening upon the heights. The first appearance of the lunar crescent was watched with care, reported by the high priest, and announced to the people by the sound of trumpets. The new moons which correspond with the renewal of the four seasons were the most solemn; we find here the origin of the "ember weeks" of the Church, as we find that of most of our festivals in the ceremonies of the ancients. The Orientals, the Chaldeans, Egyptians, and Jews religiously observed this custom.

An eclipse of the sun has already been described. An eclipse of the moon is an equally simple matter. An eclipse of the sun is caused by the dark, solid body of the moon passing just between earth and sun, hiding the sun from us, and casting its shadow upon the earth. An eclipse of the moon is also caused by a shadow—the shadow of our own earth—falling upon the moon.

Here again, if the plane of the moon's orbit were the same as ours, eclipses of the moon would be very common. As it is, her orbit carries her often just a little too high or too low to be eclipsed, and it is only now and then, at regular intervals, that she passes through the shadow of the earth.

If a large, solid ball is hung up in the air, with bright sunlight shining on it, the sunlight will cast a *cone of shadow* behind the ball. It will throw, in a direction just away from the sun, a long, round shadow, the same as the ball at first, but tapering gradually off to a point. If the ball is near the ground, a round shadow will rest there, almost as large as the ball. The higher the ball is placed, the smaller will be the round shadow, till at length, if the ball be taken far enough upwards, the shadow will not reach the ground at all. Our earth and all the planets cast just such tapering cones of dark shadow behind them into space. The cone always lies in a direction away from the sun.

It is when the moon comes into this shadow that an "eclipse of the moon" takes place. Sometimes she only dips half-way into it, or just grazes along the edge of it, and that is called a "partial eclipse." Sometimes she goes in altogether, straight through the midst of the shadow, so that the whole of her bright face for a short time grows quite dark. Then we have a "total lunar eclipse."

CHAPTER XV.

YET MORE ABOUT THE MOON.

THERE are two ways of thinking about the moon. One way is to consider her as merely the earth's attendant satellite. The other way is to consider her as our sister-planet, traveling with us round the central sun.

The first is the more common view; but the second is just as true as the first.

For the sun does actually pull the moon towards himself, with a very much stronger pulling than that of the earth. The attraction of the sun for the moon is more than double the attraction of the earth for the moon. If it were not that he pulls the earth quite as hard as he pulls the moon, he would soon overpower the earth's attraction, and drag the moon away from us altogether.

People are often puzzled about the orbit or pathway of the moon through the heavens. For in one sense they have to think of her as traveling round and round in a fixed orbit, with the earth in the center. In another sense they have to think of her as always journeying onwards with the earth in her journey round the sun, and thus never returning to the same point.

There are two ways of meeting this difficulty. First of all, remember that the one movement does not interfere with the other. Just as in the case of the earth traveling round the sun, and also traveling onward with him through space; just as in the case of a boy walking round and round a mast, and also being borne onwards by the moving vessel,—so it is here. The two movements are quite separate and independent of each other. As regards the earth alone, the moon journeys round and round perpetually, not in a circle, but in a pathway which comes near being an ellipse. As regards the actual *line* which the moon's movements may be supposed to draw in space, it has nothing elliptical about it, since no one point of it is ever reached a second time by the moon.

But according to this last view of the question, nobody ever can or will walk in a circle or an oval. Take a walk round your grass-plot, measuring your distance carefully at all points from the center. Is that a circle? All the while you moved, the surface of the earth was rushing along and bearing you with it, and the whole earth was hurrying round the sun, and was being also carried by him in a third direction. Whatever point in space you occupied when you started, you can *never fill that particular part of space again.* The two ends of your so-called circle can never be joined.

But then you may come back to the same point *on the grass*, as that from which you started. And this is all that really signifies. Practically you have walked in a circle. Though not a circle as regards space generally, it is a circle as regards the earth. So also the moon comes back to the same point *in her orbit round the earth.* Letting alone the question of space, and considering only the earth, the moon has—roughly speaking—journeyed in an ellipse. You may, however, look at this matter in quite another light. Forget about the moon being the earth's satellite, and think of earth and moon as two sister-planets going round the sun in company.

The earth, it is true, attracts the moon. So, also, the moon attracts the earth; though the far greater weight of the earth makes her attraction to be far greater. If earth and moon were of the same size, they would pull each other with equal force.

But though the pull of the earth upon the moon is strong, the pull of the sun upon the moon is more than twice as strong. And greatly as the earth influences the moon, yet the actual center of the moon's orbit is the sun, and not the earth. Just as the earth travels round the sun, so also the moon travels round the sun.

The earth travels steadily in her path, being only a little swayed and disturbed by the attraction of the moon. The moon on the contrary, while traveling in her orbit, is very much swayed and disturbed indeed by the earth's attraction. In fact, instead of being able to journey straight onwards like the earth, her orbit is made up of a succession of delicate curves or scallops, passing alternately backwards and forwards over the orbit of the earth. Now she is behind the earth; now in front of the earth; now between earth and sun; now outside the earth away from the sun. The order of positions is not as here given, but each is occupied by her in turn. Sometimes she moves quickly, sometimes she moves slowly, just according

to whether the earth is pulling her on or holding her back. Two hundred and forty thousand miles sounds a good deal. That is the distance between earth and moon. But it is, after all, a mere nothing, compared with the ninety-three millions of miles which separate the sun from the earth and moon.

If we made a small model, with the sun in the center, and the earth and moon traveling a few inches off, only one slender piece of wire would be needed to represent the path of earth and moon together. For not only would the earth and the moon be so small as to be quite invisible, but the whole of the moon's orbit would have disappeared into the thickness of the single wire. This question of the moon's motions is in its nature intricate, and in its details quite beyond the grasp of any beginner in astronomy. But so much at least may be understood, that though the earth's attraction powerfully affects the moon, and causes in her motions *perturbations*, such as have been already spoken about as taking place among the planets, yet that in reality the great controlling power over the moon is the attraction of the sun.

The tides of the ocean are chiefly brought about by the moon's attraction. The sun has something to do with the matter, but the moon is the chief agent. This action of the moon can best be seen in the southern hemisphere, where there is less land. As the moon travels slowly round the earth, her attraction draws up the yielding waters of the ocean in a vast wave which travels slowly along with her. The same pulling which thus lifts a wave on the side of the earth towards the moon, also pulls the earth gently

away from the water on the opposite side, and causes a second wave there. The parts of the ocean between these two huge waves are depressed, or lower in level. These two waves on the opposite sides of the earth sweep steadily onwards, following the moon's movements,—not real, but seeming movements, caused by the turning of the earth upon her axis.

Once in every twenty-four hours these wide waves sweep round the whole earth in the southern ocean. They can not do the same in the north, on account of the large continents, but offshoots from the south waves travel northwards, bringing high-tide into every sea and ocean inlet. If there were only one wave, there would be only one tide in each twenty-four hours. As there are two waves, there are two tides, one twelve hours after the other. In the space between these two high-tides we have low-tide.

Twice every month we have very high and very low tides. Twice every month we have tides not so high or so low. The highest are called "spring-tides," and the lowest "neap-tides." When the moon is between us and the sun, or when she is "new moon," there are spring-tides; for the pull or attraction of sun and moon upon the ocean act exactly together. It is the same at full moon, when once more the moon is in a straight line with earth and sun. But at the first and last quarters, when the moon has her *sideways* position, and when the sun pulls in one direction and the moon pulls in another, each undoes a little of the other's work. Then we only have neap-tides; for the wave raised is smaller, and the water does not flow so high upon our shores.

In speaking of the surface of the moon, we are able only to speak about one side. The other is entirely hidden from us. This is caused by the curious fact that the moon turns on her axis and travels round the earth in exactly the same length of time. One-half of the moon is thus always turned towards us, though of that half we can only see so much as is receiving the light of the sun. But the half turned in our direction is always the same half. One part of the moon—not quite so much as half, though always the same portion—is turned away from us. A small border on each side of that part becomes now and then visible to us, owing to certain movements of the earth and the moon.

What sort of a landscape may lie in the unknown district, it is idle to imagine. Many guesses have been made. Some have supposed it possible that air *might* be found there; that water *might* exist there; that something like earthly animals *might* live there. It is difficult to say what may not be, in a place about which we know nothing whatever. But judging from our earthly experience, nothing seems more unlikely than that air, water, clouds, should be entirely banished from over one-half of a globe, and collected together in the space remaining.

We are on safer ground when we speak about that part of the moon which is turned towards us. For we can say with confidence that if any atmosphere exist there, it must be in thickness less than the two-thousandth part of our earthly atmosphere. It seems equally clear that water also must be entirely wanting. The tremendous heat of the long lunar day would

raise clouds of vapor, which could not fail to be visible. But no such mistiness ever disturbs the sharply-defined outline of the moon, and no signs of water action are seen in the craggy mountains and deep craters.

The craters which honeycomb the surface of the

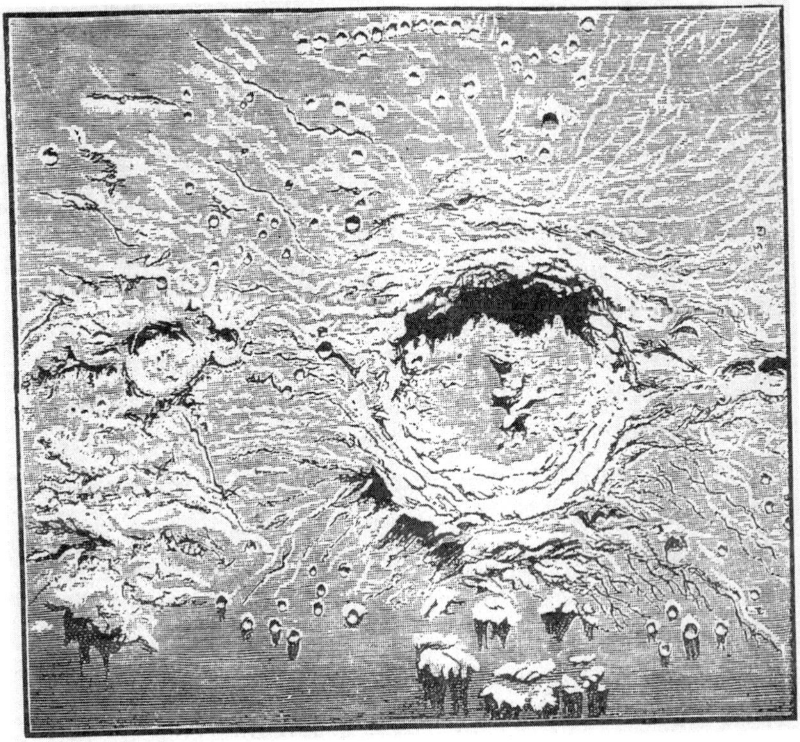

THE LUNAR CRATER COPERNICUS.

moon are various in size. Many of the larger ones are from fifty to a hundred miles in diameter. These huge craters—or, as we may call them, deep circular plains—are surrounded by mighty mountain ramparts, rising to the height of thousands of feet. Usually they have in their center a sugar-loaf or

cone-shape mountain, or even two or more such mountains, somewhat lower in height than the surrounding range. The sunset-lights upon certain of these distant mountain-peaks were first watched by Galileo through his telescope, and have since been seen by many an observer—intense brightness contrasting with intense blackness of shadow.

In addition to her great craters, the moon seems to be thickly covered with little ones, many of them being as small as can be seen at all through a telescope. Whether these are all volcano-craters remains to be discovered. It is not supposed that any of them are now active. From time to time, signs of faint changes on the moon's surface have been noticed, which it was thought might be owing to volcanic outbursts. Such an outburst as the worst eruptions of Mount Vesuvius would be invisible at this distance. But the said changes may be quite as well accounted for by the startling fortnightly variations of climate which the moon has to endure. The general belief now inclines to the idea that the moon-volcanoes are extinct, though no doubt there was in the past great volcanic activity there.

A description has been given earlier of the rain of meteorites constantly falling to our earth, and only prevented by the atmosphere from becoming serious. But the moon has no such protecting atmosphere, and the amount of cannonading which she has to endure must be by no means small. Perhaps in past times, when her slowly-cooling crust was yet soft, these celestial missiles showering upon her may have occasionally made deep round holes in her surface.

This is another guess, which time may prove to be true. Guesses at possible explanations of mysteries do no harm, so long as we do not accept them for truth without ample reason.

The origin of the lunar craters must be referred to some ancient epoch in the moon's history. How ancient that epoch is we have no means of knowing;

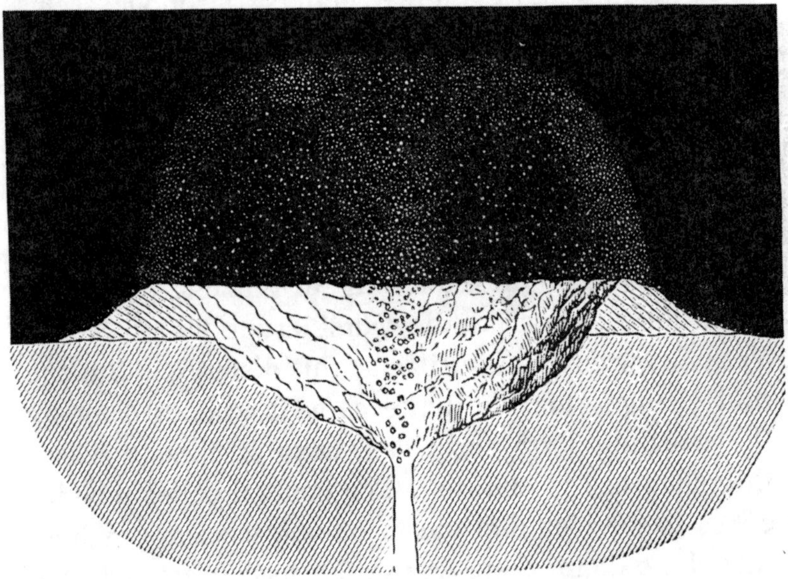

but in all probability the antiquity of the lunar craters is enormously great. At the time when the moon was sufficiently heated to have these vast volcanic eruptions, of which the mighty craters are the survivals, the earth must have been very much hotter than it is at present. It is not, indeed, at all unreasonable to believe that when the moon was hot enough for its volcanoes to be active, the earth was so hot that life was impossible on its surface. This supposition would point to an antiquity for the moon's craters far too great to be estimated by the centuries and the thousands of years which are adequate for the lapse of time as recognized by the history of human events. It seems not unlikely that millions of years may have elapsed since the mighty craters of Plato or of Copernicus consolidated into their present form.

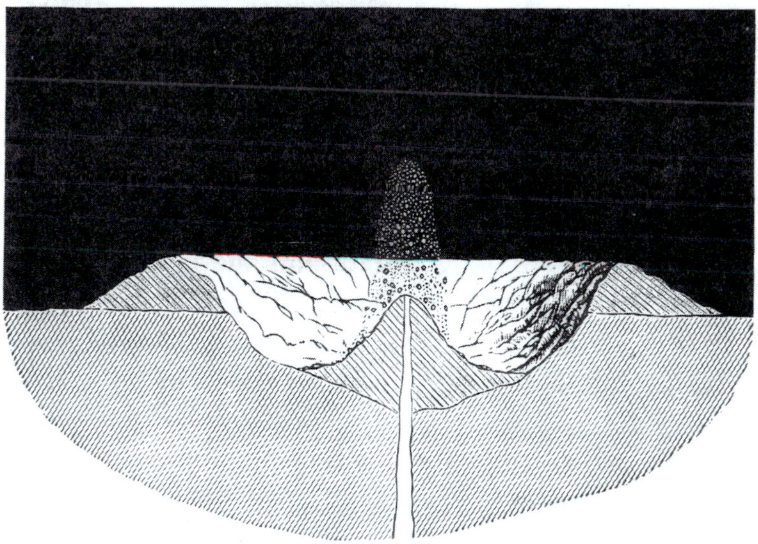

LUNAR ERUPTION—FEEBLE ACTION.

It will now be possible for us to attempt to account for the formation of the lunar craters. The most probable views on the subject are certainly those adopted by Mr. Nasmyth, as represented in the cuts, though it must be admitted that they are by no means free from difficulty. We can explain the way in which the rampart around the lunar crater is formed, and the great mountain which so often adorns the center of the plain. The first of these cuts contains an imaginary sketch of a volcanic vent on the moon in the days when the craters were active. The eruption is here in the full flush of its energy, when the internal forces are hurling forth a fountain of ashes or stones, which fall at a considerable distance from the vent; and these accumulations constitute the rampart surrounding the crater. The second cut depicts the crater in a later stage of its history. The prodigious explosive power has now been exhausted, and perhaps has been intermitted for some time. A feeble jet issues from the vent, and deposits the materials close around the orifice, and thus gradually raises a mountain in the center.

Besides the craters and their surrounding barriers, there are ranges of mountains on the moon, and flat plains which were once named "seas," before it was found that water did not exist there. Astronomers also see bright ridges, or lines, or cracks of light, hard to explain.

One of the chief craters is called "Ptolemy," and in size it is roughly calculated to be no less than one hundred and fourteen miles across. Another, "Copernicus," is about fifty-six miles; and another, "Tycho," about fifty-four miles. The central cone-mountain of Tycho is five thousand feet high. The crater of "Schickard" is supposed to be as much as one hundred and thirty-three miles in diameter.

Astronomers have agreed to name these craters after the great discoverers who enlarged our knowledge of the solar and sidereal systems. It is fitting that these great names are suggested every time the moon is seen through a telescope. To Ptolemy we are indebted for what is known as "The Ptolemaic System of the Universe," which makes the earth the center around which the sun, moon, planets, and stars all revolve, and explains the apparently erratic movements of the planets by supposing their orbits to be epicycles; that is, curves returning upon themselves and forming loops. Tycho Brahe was willing to allow, with Copernicus, that the planets all revolved around the sun; but he taught that both sun and planets turned around the earth.

The system known as "The Copernican" is now known to be the only true one, and it is universally accepted. It is so called after Nicolaus Copernicus, who was born in Thorn, Prussia, February 19, 1473. He was educated for the Church, and studied medicine, but devoted himself especially to astronomy.

Without being precisely a great genius, this quiet and thoughtful monk seems to have been wise far beyond the age in which he lived, and remarkable for his independence of mind. He had a "profound sagacity," and a wide general grasp of scientific subjects.

The extremely complicated and cumbrous nature of the Ptolemaic system appeared to his judgment hardly compatible with the harmony and simplicity elsewhere characteristic of nature. Moreover, he was impressed and perplexed by the very marked changes in the brilliancy of the planets at different seasons. These changes *now* are no difficulty at all. Venus, Mars, Jupiter, when on the same side of the sun as ourselves, are comparatively near to us, and naturally look much more bright in consequence of that nearness than when they are on the opposite side of the sun from ourselves. But under the Ptolemaic system, each planet was supposed to revolve round our earth, and to be always at about the same distance from us; therefore, why such variations in their brilliancy?

NICOLAUS COPERNICUS.

During thirty-six long years he patiently worked out this theory, and during part of those thirty-six years he wrote the one great book of his lifetime, explaining the newer view of the Solar System which had taken hold of his reason and imagination. This book, named *De Revolutionibus Orbium Cœlestium*, or, "Concerning the Revolutions of the Celestial Spheres," which came out only a few hours before his death, was dedicated to the Bishop of Rome—a little touch of worldly wisdom which doubtless staved off for a while the opposition of the Vatican.

Copernicus was not the first who thought of interpreting the celestial motions by the theory of the earth's motion. That immortal astronomer has taken care to give, with rare sincerity, the passages in the ancient writers from which he derived the first idea of the probability of this motion—especially Cicero, who attributed this opinion to Nicetas of Syracuse; Plutarch, who puts forward the names of Philolaus, Heraclides of Pontus, and Ecphantus the Pythagorean; Martianus Capella, who adopted, with the Egyptians, the motion of Mercury and Venus around the sun, etc. Even a hundred years before the publication of the work of Copernicus, Cardinal Nicolas of Cusa, in 1444, in his great theological and scientific encyclopedia, had also spoken in favor of the idea of the earth's motion and the plurality of worlds. From ancient times to the age of Copernicus, the system of the earth's immobility had been doubted by clear-sighted minds, and that of the earth's motion was proposed under different forms. But all these attempts still leave to Copernicus the glory of establishing it definitely.

Not content with merely admitting the idea of the earth's motion as a simple arbitrary hypothesis, which several astronomers had done before him, he wished—

and this is his glory—to demonstrate it to himself by acquiring the conviction by study, and wrote his book to prove it. The true prophet of a creed, the apostle of a doctrine, the author of a theory, is the man who, by his works, demonstrates the theory, makes the creed believed in, and spreads the doctrine. He is not the creator. "There is nothing new under the sun," says an ancient proverb. We may rather say, Nothing which succeeds is entirely new. The new-born is unformed and incapable. The greatest things are born from a state of germ, so to say, and increase unperceived. Ideas fertilize each other. The sciences help each other; progress marches. Men often feel a truth, sympathize with an opinion, touch a discovery, without knowing it. The day arrives when a synthetical mind feels in some way an idea, almost ripe, becoming incarnate in his brain. He becomes enamored of it, he fondles it, he contemplates it. It grows as he regards it. He sees, grouping round it, a multitude of elements which help to support it. To him the idea becomes a doctrine. Then, like the apostles of good tidings, he becomes an evangelist, announces the truth, proves it by his works, and all recognize in him the author of the new contemplation of nature, although all know perfectly well that he has not invented the idea, and that many others before him have foreseen its grandeur.

Such is the position of Copernicus in the history of astronomy. The hypothesis of the earth's motion had been suggested long before his birth on this planet. This theory counted partisans in his time But he—he did his work. He examined it with the patience of an astronomer, the rigor of a mathematician, the sincerity of a sage, and the mind of a philosopher. He demonstrated it in his works. Then he died without seeing it understood, and it was not till a century after his death that astronomy adopted it, and popularized it by teaching it. However, Copernicus is really the author of the true system of the world, and his name will remain respected to the end of time.

The so-called "seas" on the moon are those large dark spots to be seen on its surface, in the shape of "eyes, nose, and mouth," or of the famous old man with his bundle of sticks. The brighter parts are the more mountainous parts.

The chief ranges of lunar mountains have been named by astronomers after mountains on earth, such as the Apennines, the Alps, the Caucasian range, the Carpathian and the Altai Mountains.

CHAPTER XVI.

MERCURY, VENUS, AND MARS.

ONCE again we have to journey through the high-roads of the Solar System, paying a brief visit to each in turn of our seven chief brother-and-sister planets, and learning a few more leading facts about them. Having gone the same way before, it will not now seem quite so far.

Busy, hurrying Mercury! we must meet him first in his wild rush through space. If he were to slacken speed for a single instant, he would begin to fall with fearful rapidity towards the sun. And if Mercury were to drop into one of those huge black chasms of rent furnace-flame on the sun's surface, there would be a speedy end to his life as a planet.

Mercury's day is about the same length as our day, and his year is about one quarter the length of our year. If Mercury has spring, summer, autumn, and winter, each season must be extremely short; but this depends upon whether Mercury's axis slopes like the earth's axis—a matter difficult to find out. Mercury is always so near to the sun, that it is by no means easy to observe him well.

We know more about his orbit than his axis. The earth's orbit, as before explained, is not a circle, but an ellipse or oval. Mercury's orbit is an ellipse also, and a much longer—or, as it is called, a more eccentric—ellipse. The earth is three millions of miles nearer to the sun at one time of the year, than six months before or after. Mercury is no less than fifteen millons of miles nearer at one time than another, which must make a marked difference in the amount of heat received.

Even when the distance is greatest, the sun as seen from Mercury looks four and a half times as large as the sun we see. What a blazing splendor of light! It is not easy to imagine human beings living there, in such heat and glare, and with either no changes of season at all, or such very short seasons rapidly following one another. Mercury may, and very likely does, abound with living creatures, as much as the earth abounds with them; only one fancies they must be altogether a different kind of living creatures from any ever seen on earth. And yet we do not know. Man can so wonderfully adapt himself or be adapted to different climates on earth, from extreme heat to extreme cold, that we can not tell how far this adapting power may reach.

Both Mercury and Venus seem to be enfolded in dense, cloud-laden atmospheres, rarely parting so as to allow us to get even a glimpse of the real planets within the thick, light-reflecting covering. Some have thought that a heavy, moist, protecting atmosphere may help to ward off the intense heat, and to make Mercury a more habitable place. Our earthly atmosphere is rather of a kind to store up heat, and to make us warmer than we should be without it; but there might be vapors differently constituted which might act in some other way. At all events, we know how easily God can have adapted either the planet to the creatures he meant to place there, or the creatures to the climate. "All things are possible" to him. The how and the what are interesting questions for us, but we must often be content to wait for an answer.

A thick, gray ring or belt has been noticed round the small, black disk of Mercury, while it has passed between us and the sun. The edge of Mercury, seen against the bright photosphere beyond, would, if there were no atmosphere, be sharp and clear as the edge of the airless moon. This surrounding haze seems to show that Mercury has an atmosphere. The sunlight reflected from Mercury's envelope of clouds shines at least as brightly as if it were reflected from his solid body.

The small size of Mercury makes attraction on his surface much less than on earth. A lump of iron weighing on earth one pound, would weigh on Mercury only about seven ounces, or less than half as much. So a man would be a very light leaper indeed there, and an elephant might be quite a frolicsome animal. If there are star-gazers in Mercury, and if the cloud-laden atmosphere allows many clear views of the sky, the earth and Venus must both be beautiful to look upon. Each of the two would shine far more brightly than Jupiter, as seen at his best from earth.

Like Mercury, Venus, the next planet, has an orbit lying inside our orbit. Mercury and Venus are always nearer to the sun than we are; and if Mercury and Venus traveled round the sun in orbits, the planes of which were exactly the same as the plane of the earth's orbit, we should very often see them creeping over the surface of the sun. Not that they really "creep over" it; only, as they journey between the sun and us, we can see them pass like little black dots across the sun's disk. This is the same thing as when the moon passes across the sun's disk and eclipses it. But Mercury and Venus are too far away from us to cause any eclipse of the sun's light.

Mercury has given more trouble to astronomers than any other member of the system; for, owing to his proximity to the sun, he is usually lost in the solar glory, and is never seen in a dark part of the heavens, even at the time of his greatest distance. This circumstance, together with a small mass and an immense velocity, renders it difficult to catch and watch him. In high latitudes, where the twilight is strong and lengthened, while mists often overhang the horizon, the planet can seldom be seen with the naked eye. Hence an old astro-meteorologist contemptuously describes him as "a squirting lacquey of the sun, who seldom shows his head in these parts, as if he was in debt." Copernicus lamented that he had never been able to obtain a sight of Mercury; and the French astronomer Delambre saw him not more than twice with the naked eye.

The most favorable times for making observations are about an hour and three-quarters before sunrise in autumn, and after sunset in spring; but very clear weather and a good eye are required. At certain periods, when between the earth and the sun, on a line joining the centers of the two bodies, Mercury appears projected on the solar disk as a small, round, dark spot. This is called a *transit*, and would occur during every revolution if the plane of his orbit coincided with that of the orbit of the earth. But as one-half of his orbit is a little above that of the earth, and the other half a little below it, he passes above or below the sun to the terrestial spectator, except at those intervals, when, being between us and the sun, he is also at one of the two opposite points where the planes of the respective orbits intersect each other. Then, stripped of all luster, the planet passes over the face of the great luminary as a black circular speck, affording evidence of his shining by reflected light, and of his spherical form.

The first transit of Mercury recorded in history was predicted by Kepler, and witnessed by Gassendi, at Paris, on the morning of a cloudy day, the 7th of November, 1631. It was toward nine o'clock when he saw the planet; but owing to its extreme smallness, he was at first inclined to think it was a minute solar spot. It was then going off the sun; and made its final egress toward half-past ten. The observed time of the transit was nearly five hours in advance of the computed time. This was a very satisfactory accordance for that age between theory and observation; but it was the effect of a fortuitous combination of circumstances, rather than of computations founded upon well-established data. "The crafty god," wrote Gassendi, in the peculiar style of his day, "had sought to deceive astronomers by passing over the sun a little earlier than was expected, and had drawn a veil of dark clouds over the earth in order to make his escape more effectual. But Apollo, knowing his knavish tricks from his infancy, would not allow him to pass altogether unnoticed. To be brief, I have been more fortunate than those hunters after Mercury who have sought the cunning god in the sun. I found him out, and saw him where no one else had hitherto seen him." Since Gassendi's time a number of transits have been observed.

These crossings of the sun's face, or "transits," as they are called, have been important matters. The transit of Venus especially was once eagerly looked for by astronomers, since, by close observations of Venus's movements and positions, the distance of the sun could at that time be better calculated than in any other way. Other methods are now coming into vogue.

The transits of Venus are rare. Two come near together, separated by only eight years, and then for more than one hundred years the little dark body of Venus is never seen from earth to glide over the sun's photosphere. There was a transit of Venus in the year 1761, and another in the year 1769. There was a transit of Venus in 1874, and another in 1882. At the last transits it was found that the sun, instead of being ninety-five millions of miles away, as astronomers thought, was only ninety-three millions of miles away. The reason why these transits happen so seldom, is that the orbits of Mercury and Venus lie in rather a different plane or level from the earth's orbit. So, like the moon, though often passing between us and the sun, they generally go just a little higher or just a little lower than his bright face.

Mercury and Venus show phases like the moon, although they do not circle round the earth as the moon does. These "phases," or changes of shape, are probably never visible except through a telescope. It will be easier to think about the phases of Venus alone, than to consider both together. Her orbit lies within the earth's orbit, and the earth and Venus travel round the sun—as do all the planets—in the same direction. But as Venus's pathway is shorter than ours, and as her speed is greater, she is much the quickest about her yearly journey, and she overtakes us again and again at different points of our orbit in turn.

PHASES OF VENUS.

At one time she comes between us and the sun. That is her nearest position to us, and she is then only about twenty-five millions of miles distant. A beautiful sight she would be, but unfortunately her bright side is entirely turned away, and only her dark side is turned towards us. So then she is "new Venus," and is invisible.

At another time she is completely beyond the sun, and at her farthest position away from us. Her shining is quite lost in the sun's rays coming between. And though we get a good view of her as "full Venus," at a little to one side or the other, yet so great is her distance—as much as one hundred and fifty-seven millions of miles—that her size and brightness are very much lessened.

Between these two nearest and farthest points, she occupies two middle distances, one on each side of the sun. Then, like the moon at her "quarters," she turns to us only half of her bright side. But this is the best view of Venus that we have, as a brilliant, untwinkling, starlike form,—the Evening Star of ancients and of poets. Between these four leading positions Venus is always traveling gradually from one to another—always either waxing or waning in size and in brightness. Mercury passes through the same seeming changes.

Inhabitants of Venus must have a glorious view of the earth, with her attendant moon. For just at the time when the two planets are nearest together, and when she is only "new Venus" to us, a dark and invisible body, the earth is "full earth" to Venus. The very best sight we ever have of Venus can not come near that sight. But if Mercury and Venus really are so often covered with heavy clouds as astronomers believe, this must greatly interfere with any habits of star-gazing.

Venus and the earth have often been called twin-sister planets. There are many points of likeness between them. In size they differ little, and in length of day they are within an hour of being the same. Earth certainly has a companion-moon, and Venus, it is believed, has not. At one time several astronomers were pretty certain that they had caught glimpses of a moon; but the supposed moon has of late quite vanished, and nobody can say whether it ever really existed. Venus travels in an ellipse which comes nearer to being a circle than the orbit of any other planet.

Mountain-shadows have been watched through the telescope, in Venus, as in the moon. Some astronomers have believed that they saw signs of very lofty mountains—as much as twenty-eight miles, or four times the height of our highest earthly mountains, but this requires confirmation.

There is a good deal of uncertainty about the climate of Venus. The heat there must greatly surpass heat ever felt on earth—the sun being about double the apparent size of our sun, and pouring out nearly double the amount of light and heat that we receive.

This difference may be met, as already stated, by a sheltering, cloudy atmosphere, or the inhabitants may have frames and eyesight suited to the increased glare and warmth.

An atmosphere and water exist there as here. From what we have seen above of the rapid and violent seasons of this planet, we might think that the agitations of the winds, the rains, and the storms would surpass everything which we see and experience here, and that its atmosphere and its seas would be subject to a continual evaporation and precipitation in torrential rains—an hypothesis confirmed by its light, due, doubtless, to reflection from its upper clouds, and to the multiplicity of the clouds themselves. To judge by our own impressions, we should be much less pleased with this country than with our own, and it is even very probable that our physical organization, accommodating and complaisant as it is, could not become acclimatized to such variations of temperature. But it is not necessary to conclude from this that Venus is uninhabitable and uninhabited. We may even suppose, without exaggeration, that its inhabitants, organized to live in the midst of these conditions, find themselves at their ease, like a fish in water, and think that our earth is too monotonous and too cold to serve as an abode for active and intelligent beings.

Of what nature are the inhabitants of Venus? Do they resemble us in physical form? Are they endowed with an intelligence analogous to ours? Do they pass their life in pleasure, as Bernardin de St. Pierre said; or, rather, are they so tormented by the inclemency of their seasons that they have no delicate perception, and are incapable of any scientific or artistic attention? These are interesting questions, to which we have no reply. All that we can say is, that organized life on Venus must be little different from terrestrial life, and that this world is one of those which resemble ours most. It should, then, be inhabited by vegetable, animal, and human races but little different from those which people our planet. As to imagining it desert or sterile—this is an hypothesis which could not arise in the brain of any naturalist. The action of the divine sun must be there, as in Mercury, still more fertile than his terrestrial work, already so wonderful. We may add that Venus and Mercury, having been formed after the earth, are relatively younger than our planet.

It is believed also that the axis of Venus, instead of being slanted only as much as the earth's axis, is tilted much more. Even if the tilting is less than some have supposed, it is probably very considerable.

If Venus really does "lie over" in such a manner, certain startling changes of climate on its surface—unpleasant changes, according to our ideas—would take place.

Like earth, Venus would have her two arctic regions, where a burning summer's day would succeed a bitter winter's night, each half a year in length. She would have also her tropical region—only in that region intense cold would alternate with intense heat, brief seasons of each in turn. And between the tropics and the arctic regions would lie wide belts, by turns entirely tropical and entirely arctic. The rapidity and severity of these changes, following one another in a year about as long as eight of our months, would seem to be too much for any human frame to endure. But it all rests upon an *if*. And we may be quite sure that *if* there are any manner of human beings in Venus, their frames are well suited to the climate of their world.

Though Mars is one of the inner group of four small planets, divided by the zone of asteroids from the outer group of four great planets, yet he belongs to the outside set of Superior Planets. His orbit surrounds ours, being at all points farther off from the sun. Very slight "phases" have been seen in Mars. He turns to us, from time to time, just enough of his dark side to prove that he *has* a dark side, and that he does not shine like a star by his own light. But the phases on that planet are by no means marked as they are with Venus.

Until lately it was believed that Mars possessed no moons. Two very small ones have, however, been lately found circling round him.* They have been named Deimos and Phobos, after the "sons of Mars" in Greek mythology. Deimos travels round Mars in thirty-nine hours, while Phobos performs the same journey in the astonishingly short period of seven hours and a half!

We have here, then, a system very different from that of the earth and moon. But the most curious point is the rapidity with which the inner satellite of

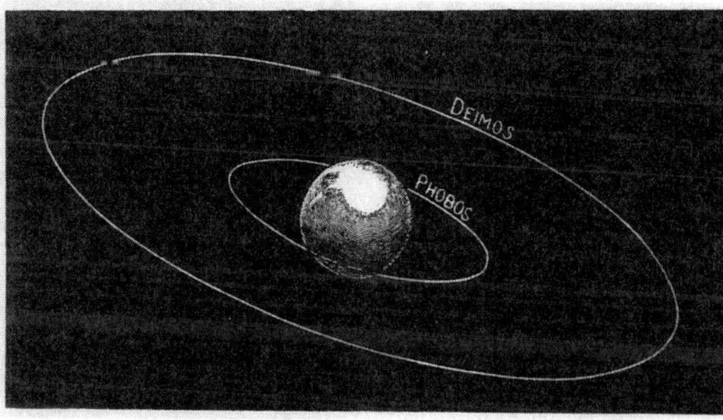

MARS AND THE PATH OF ITS SATELLITES.

Mars revolves round its planet. This revolution is performed in seven hours, thirty-nine minutes, fifteen seconds, although the world of Mars rotates on itself in twenty-four hours, thirty-seven minutes—that is to

* The discovery of the two moons is due to the agency of a woman. Professor Asaph Hall, of Washington City, was searching by the aid of the most powerful telescope which had yet been directed to Mars, and at the very moment when the planet was in the most favorable condition for observation. Having searched in vain during several evenings in August, 1877, he was about to give up, when Mrs. Hall begged him to search a little more. He did so, and on the night of the 11th he discovered the first of the satellites, and then on the 17th the second.

say, this moon turns much more quickly than the planet itself. This fact is inconsistent with all the ideas we have had up to the present on the law of formation of the celestial bodies. Thus, while the sun appears to revolve in the Martian sky in a slow journey of more than twenty-four hours, the inner moon performs its entire revolution in a third of a day. It follows that *it rises in the west* and *it sets in the east!* It passes the second moon, eclipses it from time to time, and goes through all its phases in eleven hours, each quarter not lasting even three hours. What a singular world!

These satellites are quite small—they are the smallest celestial bodies we know. The brightness of the planet prevents us from measuring them exactly. It seems, however, that the nearer is the larger, and shows the brightness of a star of the tenth magnitude, and that the second shines as a star of the twelfth magnitude. According to the most trustworthy photometric measures, the first satellite may have a diameter of 7.45 miles, and the second a diameter of 6.2 miles. *The larger of these two worlds is scarcely larger than Paris.* Should we honor them with the title of worlds? They are not even terrestrial continents, nor empires, nor kingdoms, nor provinces, nor departments. Alexander, Cæsar, Charlemagne, or Napoleon, might care but little to receive the scepter of such worlds. Gulliver might juggle with them. Who knows, however? The vanity ot men being generally in the direct ratio of their mediocrity, the microscopical reasoning mites which doubtless swarm on their surface have also, perhaps, permanent armies, which mutilate each other for the possession of a grain of sand.

These two little moons received from their discoverer the names of *Deimos* (Terror) and *Phobos* (Flight), suggested by the two verses of Homer's "Iliad" which represent Mars descending on the earth to avenge the death of his son Ascalaphus:

> "He ordered Terror and Flight to yoke his steeds,
> And he himself put on his glittering arms."

Phobos is the name of the nearer satellite; *Deimos*, that of the more distant.

The existence of these little globes had already been suspected from analogy, and thinkers had frequently suggested that, since the earth has one satellite, Mars should have two, Jupiter four, Saturn eight; and this is indeed the fact—though as Jupiter is now found to have *five* satellites, this arithmetical progression is upset. But as we experience too often in practice the insufficiency of these reasonings of purely human logic, we can not give them more value than they really possess. We might suppose in the same way now that Uranus has sixteen satellites, and Neptune thirty-two. This is possible; but we know nothing of them, and have not even the right to consider this proportion as probable. It is not the less curious to read the following passage, written by Voltaire in 1750 in his masterpiece, the "Micromégas:"

"On leaving Jupiter, our travelers crossed a space of about a hundred millions of leagues, and reached the planet Mars. They saw *two moons*, which wait on this planet, and which have escaped the gaze of astronomers. I know well that Father Castel wrote against the existence of these two moons; but I agree with those who reason from analogy. These good philosophers know how difficult it would be for Mars, which is so far from the sun, to get on with less than two moons. However this may be, our people found it so small that they feared they might not find anything to lie upon, and went on their way."

Here we have unquestionably a very clear prophecy, a rare quality in this kind of writing. The astronomico-philosophical romance of "Micromégas" has been considered as an imitation of Gulliver. Let us open the masterpiece of Swift himself, composed about 1720, and we read, word for word, in Chapter III of the "Voyage to Laputa:"

"Certain astronomers . . . spend the greatest part of their lives in observing the celestial bodies, which they do by the assistance of glasses far excelling ours in goodness. For this advantage hath enabled them to extend the discoveries much farther than our astronomers in Europe; for they have made a catalogue of ten thousand fixed stars, whereas the largest of ours do not contain above one-third part of that number. They have likewise discovered two lesser stars, or satellites, which revolve about Mars, whereof the innermost is distant from the center of the primary planet exactly three of his diameters, and the outermost five. The former revolves in the space of ten hours, and the latter in twenty-one and a half; so that the squares of their periodical times are very near in the same proportion with the cubes of their distance from the center of Mars, which evidently shows them to be governed by the same law of gravitation that influences the other heavenly bodies."

What are we to think of this double prediction of the two satellites of Mars? Indeed, the prophecies which have been made so much of in certain doctrinal arguments have not always been as clear, nor the coincidences so striking. However, it is evident that no one had ever seen these satellites before 1877, and that there was in this hit merely the capricious work of chance. We may even remark that both the English and French authors have only spoken ironically against the mathematicians, and that in 1610, Kepler, on receiving the news of the discovery of the satellites of Jupiter, wrote to his friend Wachenfals that "not only the existence of these satellites appeared to him probable, but that doubtless there might yet be found two to Mars, six or eight to Saturn, and perhaps one to Venus and Mercury." We can not, assuredly, help noticing that reasoning from analogy is here found on the right road. However this may be, this discovery truly constitutes one of the most interesting facts of contemporary astronomy.

Mars is not only much smaller than the earth, but a good deal less dense in his "make." His material is only about three-quarters as heavy as an equal amount of the earth's material. A very heavy man on earth would be a most light and active individual on Mars. Gold taken from earth to Mars would weigh there no more than tin weighs upon earth.

Mars has, it seems, an atmosphere, even as earth has. Of all the planets Mars, is the only one whose actual surface is discernible in the telescope. Mercury and Venus are so hidden by dense envelopes of clouds that the real planets within are only now and then to be dimly caught sight of. Jupiter and Saturn are so completely enwrapped in mighty masses of vapor, that we can not even be certain whether there are any solid bodies at all inside.

But Mars can be studied. Here and there, it is true, clouds sweep over the landscape, hiding from view for a little while one continent or another, one sea or another, growing, changing, melting away, as do the clouds of earth. Still, though these clouds come and go, there are other markings on the surface of Mars which do not change. Or, rather, they only change so much as the continents and oceans of earth would seem to vary, if watched from another planet, as the daily movement of earth carried them from west to east, or as they might be hidden for a while by cloud-layers coming between. It has been curiously noted that these clouds over Mars form often in the morning and evening, and are afterwards dispersed by the heat of midday. Also there seems every reason to believe that rainfalls take place in Mars as upon earth.

The red color of Mars is well known. This does not vanish in the telescope, but it is found that parts only have the red or orange hue, while other parts are dark and greenish. These are the markings which remain always the same, and they have been so closely examined that more is known about the geography of Mars than of any other world outside our own.

Mars, at his nearest point, does not draw closer to

us than forty millions of miles. At such a distance one must not speak too confidently. There are, however, many reasons for believing that the red portions are continents and that the green portions are oceans.

The spectroscope has lately shown us that water does really exist in the atmosphere of Mars—unlike the dreary, waterless moon. So we no longer doubt that the cloudlike appearances are clouds, and that rain sometimes falls on Mars. If there is rain, and if there are clouds and vapor, there are probably oceans also.

Two singular white spots are to be seen at the north and south poles, which we believe to be polar ice and snow. Somebody looking at our earth in like manner from a distance, would doubtless perceive two such white snow-spots. These two polar caps are seen to vary with the seasons. When the north pole of Mars is turned towards the sun, the white spot there grows smaller; and at the same time, the south pole of Mars being turned away from the sun, the white spot there grows larger. Again, when the south pole is towards the sun, and the north pole away from the sun, the white spot at the south is seen to be the smallest, and the white spot at the north is seen to be the largest. This is exactly what takes place in the summers and winters of our north and south poles.

The markings of Mars have been so carefully studied, that at last a map has been made of the planet—a map of a world, never less than forty millions of miles away! Names have been given to the continents and oceans—such as Dawes Continent, Herschel Continent, De La Rue Ocean, Airy Sea, Huggins Inlet, and so on.

Land and water seem to be very differently arranged on Mars from what they are on earth. Here we have about three times as much water as land, and to get from one continent to another without crossing the sea is in some cases impossible. But a traveler there might go most conveniently to and fro, hither and thither, to all parts of his world, either on land or on water, without any change. If he preferred water, he would never need to set foot on land; and if he preferred land, he would never need to enter a boat. The two are so curiously mingled together, narrow necks of land running side by side with long, narrow sea-inlets, that Atlantic and Pacific Oceans are unknown.

Some have wondered whether the reddish color of the land may be caused by grass and trees being red instead of green. Very strange if so it were. But in that case, no doubt the inhabitants of Mars would find green just as trying to their eyesight, as we should find red trying to ours.

CHAPTER XVII.

JUPITER.

PASSING at one leap over the belt of tiny asteroids, about which we know little beyond their general movement, and the size and weight of a few among them, we reach at once the giant planet Jupiter: Mighty Jupiter, hurrying ever onward, with a speed, not indeed equal to that of Mercury or of our earth, yet eighty times as rapid as the speed of a cannon-ball! Think of a huge body, equal in bulk to twelve hundred earths, equal in weight to three hundred earths, rushing ceaselessly through space, at the rate of seven hundred thousand miles a day!

Jupiter's shape is greatly flattened at the poles. He spins rapidly on his axis, once in nearly ten hours, and has therefore a five hours' day and a five hours' night. As the slope of his axis is exceedingly slight, he can boast little or no changes of season. The climate near the poles has never much of the sun's heat. In fact, all the year round the sun must shine upon Jupiter much as he shines on the earth at the equinoxes.

But the amount of light and heat received by Jupiter from the sun is only about one twenty-fifth part of that which we receive on earth; and the sun, as seen from Jupiter, can have but a small, round surface, not even one-quarter the diameter of the sun we see in the sky.

When looked at with magnifying power, the bright, starlike Jupiter grows into a broad, softly-shining disk or plate, with flattened top and bottom, and five tiny, bright moons close at hand. Sometimes one moon is on one side, and four are on the other; sometimes

GENERAL ASPECT OF JUPITER—SATELLITE AND ITS SHADOW.

two are one side and three on the other; sometimes one or more are either hidden behind Jupiter or passing in front of him. Jupiter has also curious markings on his surface, visible through a telescope. These markings often undergo changes; for Jupiter is no chill, fixed, dead world, such as the moon seems to be.

There are dark belts and bright belts, usually running in a line with the equator, from east to west.

Across the regions of the equator lies commonly a band of pearly white, with a dark band on either side of "coppery, ruddy, or even purplish" hue. Light and dark belts follow one after another, up to the north pole and down to the south pole.

When we talk of "north and south poles" in the other planets, we merely mean those poles which point towards those portions of the starry heavens which we have chosen to call "northern" and "southern." You know that all the chief planets travel round the sun in very nearly the same *plane* or flat surface that we do ourselves. That plane is called the "plane of the ecliptic." Suppose that you had an enormous sheet of cardboard, and that in the middle of this cardboard the sun were fixed, half his body being above and half below. At a little distance, fixed in like manner in the card, would be the small body of the earth, half above and half below, her axis being in a slanting position. The piece of cardboard represents what is called in the heavens the *plane of the ecliptic*—an imaginary flat surface, cutting exactly through the middle of the sun and of the earth.

If the planets all traveled in the same precise plane, they would all be fixed in the cardboard just like the earth, half the body of each above and half below. As they do not so travel, some would have to be placed a little higher, some a little lower, according to what part of their orbits they were on. This supposed cardboard "plane of the ecliptic" would divide the heavens into two halves. One half, containing the constellations of the Great Bear, the Little Bear, Cepheus, Draco, and others, would be called the Northern

Heavens. One end of the earth's axis, pointing just now nearly to the Polar Star, we name the North Pole; and all poles of planets pointing towards this northern half of the heavens, are in like manner named by us their north poles.

With regard to west and east, lay in imagination upon this cardboard plane a watch, with its face upwards; remembering that all the planets and nearly all the moons of the Solar System are said both to spin on their axes, and to travel in their orbits round the sun, *from west to east.* Note how the hands of your watch would move in such a position. The "west to east" motions of planets and moons would be in exactly the opposite direction from what the motions of the watch-hands would be.

To return to Jupiter. It is believed that these bands of color are owing to a heavy, dense atmosphere, loaded with vast masses of cloudy vapor. By the "size" of Jupiter, we really mean the size of this outside envelope of clouds. How large the solid body within may be, or whether there is any such solid body at all, we do not know. The extreme lightness of Jupiter, as compared with his great size, has caused strong doubts on this head.

The white belts are supposed to be the outer side of cloud-masses shining in the sunlight. Travelers in the Alps have seen such cloud-masses, spreading over the whole country beneath their feet, white as driven snow, and shining in the sunbeams which they were hiding from villages below; or looking like soft masses of cotton-wool, from which the mountain-peaks rose sharply here and there.

The dark spaces between seem to be rifts or breaks in the clouds. Whether, when we look at those dark spaces, we are looking at the body of Jupiter, or only at lower layers of clouds, is not known. But sometimes blacker spots show upon the dark cloud-belts, and this seems rather as if they were only lower layers of clouds, the black spots giving us peeps down into still lower and deeper layers, or else perhaps to the planet itself. These appearances remind one strongly of the sun-spots, each with its penumbra, umbra, and nucleus. Occasionally bright white spots show, instead of dark ones. It is thought that they may be caused by a violent upward rush of dense clouds of white vapor. The white spots again recall the sun and his *faculæ*.

Jupiter's bands are not fixed. Great changes go on constantly among them. Sometimes a white band will turn dark-colored, or a dark band will turn white. Sometimes few and sometimes many belts are to be seen. Sometimes a dark belt will lie slanting across the others, nearly from north to south. Once, in a single hour, an entirely new belt was seen to come into shape. Another time, two whole belts vanished in one day. The bands, in which such rapid movements are seen, are often thousands of miles in breadth. Sometimes these wide zones of clouds will remain for weeks the same. At another time a break or rift in them will be seen to journey swiftly over the surface of the planet.

The winds on earth are often destructive. A hurricane, moving at the rate of ninety miles an hour, will carry away whole buildings and level entire plantations. Such hurricanes rarely, if ever, last more than a few hours. But winds in Jupiter, judging from the movements of the clouds, often travel at the rate of one hundred and fifty miles an hour; and that, not for hours only, but for many weeks together. What manner of living beings could stand such weather may well be questioned.

Another difficulty which arises is as to the cause of these tremendous disturbances on Jupiter. Our earthly storms are brought about by the heat of the sun acting on our atmosphere. But the sun-heat which reaches Jupiter seems very far from enough to raise such vast clouds of vapor, and to bring about such prolonged and tremendous hurricanes of wind.

What if there is another cause? What if Jupiter is *not* a cooled body like our earth, but a liquid, seething, bubbling mass of fiery heat—just as we believe our earth was once upon a time, in long past ages, before her outside crust became cold enough for men and animals to live thereon? *Then*, indeed, we could understand how, instead of oceans lying on his surface, all the water of Jupiter would be driven aloft to hang in masses of steam or be condensed into vast cloud-layers. *Then* we could understand why a perpetual stir of rushing winds should disturb the planet's atmosphere.

In that case would Jupiter be a planet at all? Certainly—in the sense of obeying the sun's control. Our earth was once, we believe, a globe of melted matter, glowing with heat—and farther back still, possibly, a globe of gas. Some people are very positive about these past changes; but it is wise not to be over-positive where we can not know to a certainty what has taken place. However, Jupiter *may* have cooled down only to the liquid state, and if he goes on cooling he may, by and by, gain a solid crust like the earth.

This idea about Jupiter's hot and molten state belongs quite to late years. Certain other matters seem to bear it out, though of actual proof we have none. It is thought, for instance, that the dull, coppery red light, showing often in the dark bands, may be a red glow from the heated body within. Also it has been calculated that Jupiter gives out much more light than our earth would do, if increased to his size and moved to his place—more, in fact, than we could reasonably expect him to give out. If so, whence does he obtain the extra brightness? If he does not shine by reflected light alone, he probably shines also in some additional degree by his own light.

But what about Jupiter being inhabited? Would it in such a case be quite impossible? "Impossible" is not a word for us to use about matters where we are ignorant. We can only say that it is impossible for us to *imagine* any kind of living creatures finding a home there, if our present notions about the present state of Jupiter are correct. Then is the chief planet of the Solar System a huge, useless monument of God's power to create? Not so fast. Even as merely such a monument, he could not be useless. And even if he were put to no present use at all, it might be merely because this is a time of preparation for the future. God has his times of long and slow preparation, alike with worlds, with nations, and with individuals.

But now as to the five moons circling round Jupiter. There used to be some very pretty ideas afloat about the wonderful beauty of the moons, as seen from Jupiter, their united brilliancy so far surpassing the shining of our one poor satellite, and making up for the dim light of Jupiter's sun.

A certain little difficulty was not quite enough considered. If anybody were living on the surface of Jupiter, he would have, one is inclined to think, small

SATELLITES OF JUPITER COMPARED WITH THE EARTH AND MOON.

chance of often seeing the moons through the cloud-laden atmosphere.

The nearest of the tour larger moons to Jupiter would, it is true, appear—when visible at all—rather bigger than ours does to us; while the two next would be almost half as large, and the farthest about a quarter as large—supposing inhabitants of Jupiter to have our powers of vision.

All taken together they would cover a considerably larger space in the sky than does our moon. But it must be remembered that Jupiter's moons, like ours, shine merely by reflected sunlight. And so dim is the sunshine at that distance compared with what it is at our distance, that all the five moons together, even if full at the same time, could only give about one-sixteenth part of the light which we obtain from our one full moon.

Besides, they never are full together, seen from any one part of Jupiter. The four inner moons are never to be seen "full" at all; for just when they might be so, they are eclipsed or shaded by Jupiter's shadow. The fourth sometimes escapes this eclipse, from being so much farther away.

A thought has been lately put forward, which may or may not have truth in it. What if—instead of Jupiter being a world, inhabited by animals and people, as is often supposed, with a small distant sun and five dim moons to give them light—what if Jupiter is himself in some sort a second sun to his moons, and what if those "moons" are really inhabited planets? It may be so. That is all we can say. The idea is not an impossible one.

The so-called "moons" are certainly small. But they are by no means too small for such a purpose. Jupiter would in that case, with his five moons circling round him, enjoying his light and warmth, be a small picture of the sun, with his four inner planets and the asteroids circling round him, basking in a more lavish amount of the same.

Picturing the moons as giving light to Jupiter, we find them seemingly dim and weak for such a purpose—though, of course, we may here make a grand mistake in supposing the eyesight of living creatures in Jupiter to be no better than our own eyesight.

Even upon earth a cat can see plainly where a man has to grope his way in darkness.

But by picturing the moons as inhabited, and Jupiter as giving out some measure of heat and light

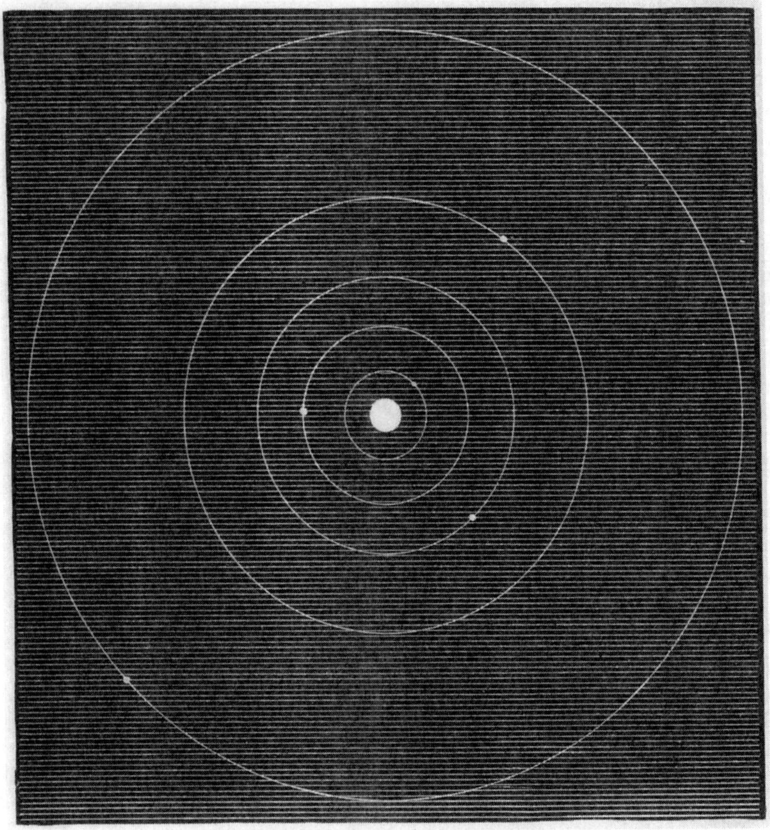

THE SYSTEM OF JUPITER.

to make up for the lessened amount of light and heat received from the sun, the matter becomes more easy to our understanding.

The nearest moon has indeed a magnificent view of Jupiter as a huge bright disk in its sky, no less than three thousand times as large as our moon appears to us, shining brightly with reflected sunlight, and it may be glowing with a red light of his own in addition. Even the farthest off of the five sees him with a face sixty-five times the size of our moon. And the varying colors and stormy changes in the cloud-belts, viewed thus near at hand, must afford marvelously beautiful effects.

Just as Mercury, Venus, Earth, and Mars travel round the sun, at different distances, nearly in the same plane, so Jupiter's five moons travel round him, at different distances, nearly in the same plane. Jupiter's moons are always to be seen in a line, not one high and another low, one near his pole and another near his equator.

The moon nearest to Jupiter, discovered in September, 1892, by Professor Barnard, now of the Yerkes Observatory of the Chicago University, has a probable diameter of one hundred miles, and revolves round its primary in about twelve hours. The second satellite, named Io, is said to be over two thousand miles in diameter, travels round Jupiter in less than two of our days, and is eclipsed by Jupiter's shadow once in every forty-two hours.

The third moon, Europa, is rather smaller, takes over three days to its journey, and suffers eclipse once in every eighty-five hours.

The fourth moon, Ganymede, is believed to be considerably larger than Mercury, journeys round Jupiter once a week, and is eclipsed once every hundred and seventy-one hours.

The fifth moon, Callisto, is also said to be slightly

larger than Mercury, performs its journey in something more than sixteen days, and from its greater distance suffers eclipse less often than the other four.

The distance of the nearest is more than one hundred thousand miles from Jupiter; that of the farthest, more than one million miles.

The following table presents more minutely the results of the latest discoveries concerning these satellites:

SATELLITES.	DISTANCE FROM CENTER OF JUPITER.	DIAMETER.	PERIOD OF REVOLUTION. D. H. M. S.
1. Barnard's Satellite,	112,500 miles.	100 miles.	11 57 23
2. Io, ...	266,000 "	2,356 "	1 18 27 33
3. Europa, ...	424,000 "	2,046 "	3 13 13 42
4. Ganymede, ...	676,000 "	3,596 "	7 3 42 33
5. Callisto, ...	1,189,000 "	2,728 "	16 15 32 11

The fact of these eclipses, and of the shadow thrown by Jupiter's body, shows plainly that though he may give out some measure of light, as has been suggested, yet that light can not be strong, or it would prevent any shadow from being thrown by the sunlight. Also, the dense masses of cloud around him, though reflecting sunlight brightly, would shut in much of his own light. Possibly it is chiefly as heat-giver and as sunlight-reflector that Jupiter serves his five satellites.

As with our own moon, so with Jupiter's moons, the real center of their orbit is the sun, and not Jupiter. They accompany Jupiter in his journey, controlled by the sun, and immensely influenced by Jupiter.

At night the spectacle of the sky seen from Jupiter is, with reference to the constellations, the same as that which we see from the earth. There, as here, shine Orion, the Great Bear, Pegasus, Andromeda, Gemini, and all the other constellations, as well as the diamonds of our sky: Sirius, Vega, Capella, Procyon, Rigel, and their rivals. The 390,000,000 of miles which separate us from Jupiter *in no way* alter the celestial perspectives. But the most curious character of this sky is unquestionably the spectacle of the five moons, each of which shows a different motion. The second moves in the firmament with an enormous velocity, and Barnard's satellite still faster, and produce almost every day total eclipses of the sun in the equatorial regions. The four inner moons are eclipsed at each revolution, just at the hours when they are at their "full." The fifth alone attains the full phase.

Contrary to the generally received opinion, these bodies do not give to Jupiter all the light which is supposed. We might think, in fact, as has been so often stated, that these five moons illuminate the nights five times better relatively than our single moon does in this respect, and that they supplement in some measure the feebleness of the light received from the sun. This result would be, assuredly, very agreeable, but nature has not so arranged it. The five satellites cover, it is true, an area of the sky greater than our moon, but they reflect the light of a sun twenty-seven times smaller than ours; indeed, the total light reflected is only equal to a sixteenth of that of our full moon, even supposing the soil of these satellites to be as white as it appears to be, especially the fifth satellite.

Jupiter appears to be a world still in process of formation, which lately—some thousands of centuries ago—served as a sun to his own system of five or perhaps more worlds. If the central body is not at present inhabited, his satellites may be. In this case, the magnificence of the spectacle presented by Jupiter himself to the inhabitants of the satellites is worthy of our attention. Seen from Barnard's satellite, Jupiter's disk has a diameter of 47°, or more than half the distance from the horizon to the zenith. Seen from the second satellite, the Jovian globe presents an immense disk of twenty degrees in diameter, or 1,400 times larger than the full moon! What a body! What a picture, with its belts, its cloud motions, and its glowing coloration, seen from so near! What a nocturnal sun!—still warm, perhaps. Add to this the aspect of the satellites themselves seen from each other, and you have a spectacle of which no terrestrial night can give an idea.

Such is the world of Jupiter from the double point of view of its vital organization and of the spectacle of external nature, seen from this immense observatory.

The attraction of the planets has always played an important part in the motion of comets and the form of their orbits. The enormous size of Jupiter gives it more influence than any other planet, and we are not surprised that it should have seriously interfered with some of the comets that belong to the Solar System. We have already seen that Biela's comet was captured by Jupiter, and that it was probably dissolved into meteoric dust. We show in the cut the orbits of eight others that circle around the sun, but retreat no farther away than Jupiter. These are the

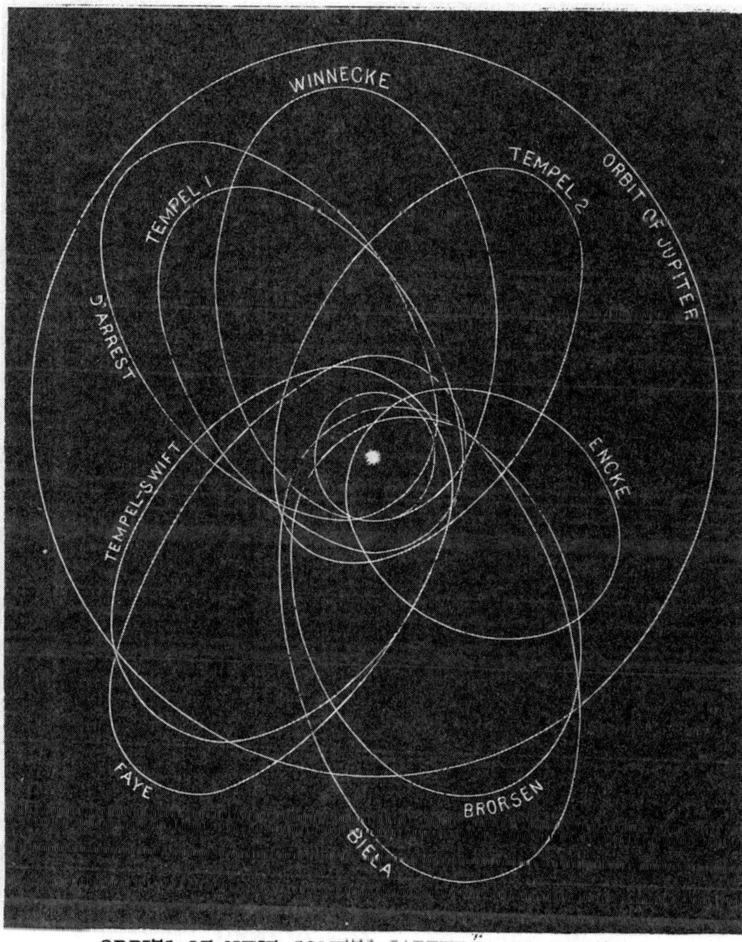

ORBITS OF NINE COMETS CAPTURED BY JUPITER.

orbits as now determined, but they vary from age to age on account of disturbances of other planets. But it is not likely that the comets themselves will ever escape from the control of Jupiter.

CHAPTER XVIII.

SATURN.

THE system of Jupiter is a simple system compared with that of Saturn, next in order. For whereas Jupiter has only five moons, Saturn has eight, and, in addition to these, he has three wonderful rings. Neither rings nor moons can be seen without a telescope, on account of Saturn's great distance from us—more than three thousand times the distance of the moon, or upwards of eight hundred millions of miles. Saturn's mean distance from the sun is eight hundred and seventy-six million seven hundred and sixty-seven thousand miles.

Saturn does not equal his mighty brother Jupiter in size, though he comes near enough in this respect to be called often his "twin"—just as that small pair of worlds, Venus and Earth, are called "twins." While Jupiter is equal in size to over one thousand two hundred earths, Saturn is equal to about seven hundred earths. And while Jupiter is equal in weight to three hundred earths, Saturn is only equal in weight to ninety earths. He appears to be made of very light materials—not more than three-quarters as dense as water. This would show the present state of Saturn to be very different from the present state of the earth. We are under the same uncertainty in speaking of Saturn as in speaking of Jupiter. Like Jupiter, Saturn is covered with dense masses of varying clouds, occasionally opening and allowing the astronomer peeps into lower cloud-levels; but rarely or never permitting the actual body of the planet to be seen.

The same perplexities also come in here, to be answered much in the same manner. We should certainly expect that in a vast globe like Saturn the strong force of attraction would bind the whole into a

SATURN AND THE EARTH—COMPARATIVE SIZE.

dense solid mass; instead of which, Saturn is about the least solid of all the planets. He seems to be made up of a light, watery substance, surrounded by vapor.

One explanation can be offered. What if the globe of Saturn be still in a red-hot, molten state, keeping such water as would otherwise lie in oceans on his surface, floating aloft in masses of steam, the outer parts of which condense into clouds?

No one supposes that Jupiter and Saturn are in the same condition of fierce and tempestuous heat as the sun. They may have been so once, but they must now have cooled down very many stages from that condition. Though no longer, however, a mass of far-reaching flames and fiery cyclones, the body of each may have only so far cooled as to have reached a stage of glowing molten red-heat, keeping all water in the form of vapor, and sending up strong rushes of burning air to cause the hurricanes which sweep to and fro the vast cloud-masses overhead.

And if this be the case, then, with Saturn as with Jupiter, comes the question, Can Saturn be inhabited? And if—though we may not say it is impossible, yet we feel it to be utterly unlikely—then again follows the question, What if Saturn's *moons* are inhabited?

Telescopic observations tempt us to believe that there is on this planet a quantity of heat greater than that which results from its distance from the sun; for the day-star as seen from Saturn is, as we have said, ninety times smaller in surface, and its heat and light are reduced in the same proportion. Water could only exist in the solid state of ice, and the vapor of water could not be produced so as to form clouds similar to ours. Now, meteorological variations are observed similar to those which we have noticed on Jupiter, but less intense. Facts, then, combine with theory to show us that the world of Saturn is at a temperature at least as high as ours, if not higher.

But the strangest feature of the Saturnian calendar is, unquestionably, its being complicated, not only with the fabulous number of 25,060 days in a year, but, further, with eight different kinds of months, of which the length varies from 22 hours to 79 days—

that is to say, from about two Saturnian days to 167. It is as if we had here *eight moons revolving in eight different periods.*

The inhabitants of such a world must assuredly differ strangely from us from all points of view. The specific lightness of the Saturnian substances and the density of the atmosphere will have conducted the vital organization in an extra-terrestrial direction, and the manifestations of life will be produced and developed under unimaginable forms. To suppose that there is nothing fixed, that the planet itself is but a skeleton, that the surface is liquid, that the living beings are gelatinous—in a word, that all is unstable—would be to surpass the limits of scientific induction.

The diameter of the largest moon is about half the diameter of the earth, or much larger than Mercury. The four inner satellites are all nearer to Saturn than our moon to us, though the most distant of the eight is ten times as far away. The inner moon takes less than twenty-three hours to travel round Saturn, and the outer one over seventy-nine days.

A great many charming descriptions have been worked up, with Saturn as with Jupiter, respecting the magnificent appearance of the eight radiant moons, joined to the glorious shining of the rings, as quite making up for the diminished light and heat of the sun. But here again comes in the doubt, whether really it is the moons who make up to Saturn for lack of light, or whether it is Saturn who makes up to the moons for lack of light.

Certainly, Saturn's cloudy covering would a little

interfere with observations of the moons by any inhabitants of the solid body within—supposing there be any solid body at all. And though it sounds very wonderful to have eight moons instead of one moon, yet all the eight together give Saturn only a very small part of the light which we receive from our one full moon—so much more dimly does the sun light them up at that enormous distance.

The same thing has been noticed with Saturn as with Jupiter—that he seems to shine more brightly than is to be expected in his position, from mere reflection of the sun's rays. A glowing body within, sending a certain amount of added light through or between the masses of clouds, would explain away this difficulty.

One more possible proof of Saturn's half-liquid state is to be found in his occasional very odd changes of shape. Astronomers have been startled by a peculiar bulging out on one side, taking off from his roundness, and giving a square-shouldered aspect. We may not say it is quite impossible that a solid globe should undergo such tremendous upheavals and outbursts as to raise a great portion of its surface five or six hundred miles above the usual level—the change being visible at a distance of eight hundred millions of miles. But it would be easier to understand the possibility of such an event, in the case of a liquid, seething mass, than in the case of a solid ball.

On the other hand this alteration of outline may be caused simply by a great upheaval not of the planet's surface, but of the overhanging layers of clouds.

Some such changes, only much slighter, have been remarked in Jupiter.

And now as to the rings. Nothing like them is to be seen elsewhere in the Solar System. They are believed to be three in number; though some would divide them into more than three. Passing completely round the whole body of Saturn, they rise, one beyond another, to a height of many thousands of miles.

The inner edge of the inner ring—an edge perhaps one hundred miles in thickness—is more than ten thousand miles from the surface of Saturn or more strictly speaking, from the outer surface of Saturn's cloudy envelope. A man standing exactly on the equator and looking up, even if no clouds came between, would scarcely be able to see such a slender dark line at such a height.

This dark, transparent ring, described sometimes as dusky, sometimes as richly purple, rises upwards to a height or breadth of nine thousand miles. Closely following it is a ring more brilliant than Saturn himself, over eighteen thousand miles in breadth. When astronomers talk of the "breadth" of these rings, it must be understood that they mean the width of the band measured *upwards*, in a direction away from the planet.

Beyond the broad, bright ring is a gap of about one thousand seven hundred miles. Then follows the third ring, ten thousand miles in breadth; its outermost edge being at a height of more than forty-eight thousand miles from Saturn. The color of the third ring is grayish, much like the gray markings often seen on Saturn.

What would not be our admiration, our astonishment, our stupor perhaps, if it were granted us to be transported there alive, and, among all these extra-terrestrial spectacles, to contemplate the strange aspect of the rings, which stretch across the sky like a bridge suspended in the heights of the firmament! Suppose we lived on the Saturnian equator itself, these rings would appear to us as a thin line drawn across the sky above our heads, and passing exactly through the zenith, rising from the east and increasing in width, then descending to the west and diminishing according to perspective. Only there have we the rings precisely in the zenith. The traveler who journeys from the equator towards either pole leaves the plane of the rings, and these sink imperceptibly, at the same time that the two extremities cease to appear diametrically opposite, and by degrees approach each other. What an amazing effect would be produced by this gigantic arch, which springs from the horizon and spans the sky! The celestial arch diminishes in height as we approach the pole. When we reach the sixty-third degree of latitude the summit of the arch has descended to the level of our horizon and the marvelous system disappears from the sky; so that the inhabitants of those regions know nothing of it, and find themselves in a less favorable position to study their own world than we, who are nearly 800,000,000 miles distant.

During one-half of the Saturnian year the rings afford an admirable moonlight on one hemisphere of the planet, and during the other halt they illuminate the other hemisphere; but there is always a half-year without "ringlight," since the sun illuminates but one face at a time. Notwithstanding their volume and number, the satellites do not give as much nocturnal light as might be supposed; for they receive, on an equal surface, only the ninetieth part of the solar light which our moon receives. All the Saturnian satellites which can be at the same time above the horizon and as near as possible to the full phase do not afford more than the hundredth part of our lunar light. But the result may be nearly the same, for the optic nerve of the Saturnians may be ninety times more sensitive than ours.

But there are further strange features in this system. The rings are so wide that their shadow extends over the greater part of the mean latitudes. During fifteen years the sun is to the south of the rings, and for fifteen years it is to the north. The countries of Saturn's world which have the latitude of Paris endure this shadow for more than five years. At the equator the eclipse is shorter, and is only renewed every fifteen years; but there are every night, so to say, eclipses of the Saturnian moons by the rings and by themselves. In the circumpolar regions the day-star is never eclipsed by the rings; but the satellites revolve in a spiral, describing fantastic rounds, and the sun himself disappears at the pole during a long night of fifteen years.

At one time it was supposed that the rings were solid, but they are now believed to consist of countless myriads of meteorites, each whirling in its own appointed pathway round the monster planet.

As already said—leaving out of the question the

cloudy atmosphere—a man standing on the equator would see nothing of the rings. A man standing at the north pole or the south pole of Saturn could see nothing either, since the rings would all lie below his horizon. But if he traveled southward from the north

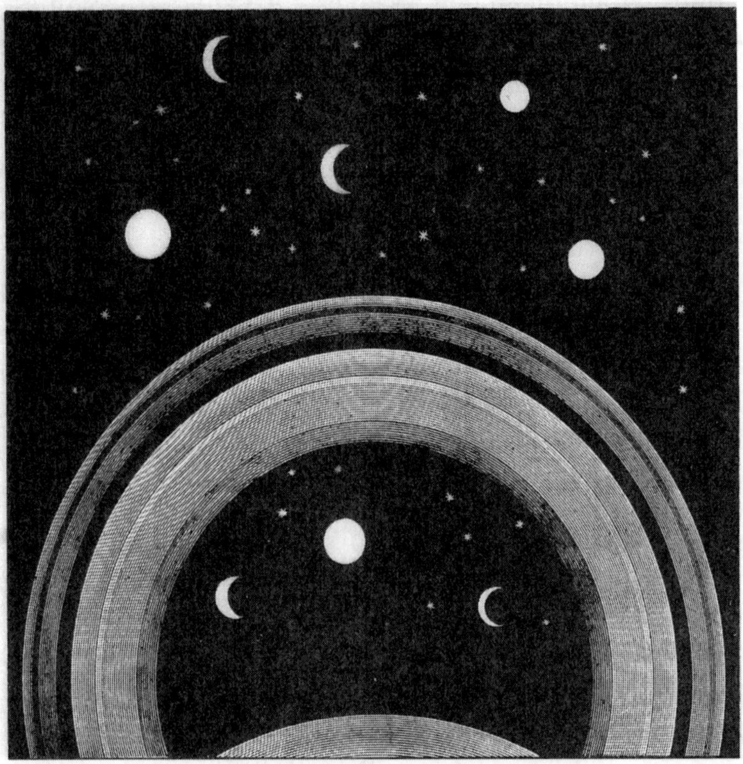

IDEAL VIEW OF SATURN'S RINGS AND SATELLITES FROM THE PLANET.

pole, or northward from the south pole, towards the equator, he would in time see the ringed arch appearing above the horizon, rising higher and growing wider with every mile of his journey; and when he was in a position to view the whole broad expanse, the transparent half-dark belt below, the wide radiant band rising upwards over that, and the grayish border surmounting all, he would truly have a magnificent spectacle before him.

This magnificent spectacle is, however, by no means always visible, even from those parts of Saturn where alone it ever can be seen. The rings shine merely by reflected sunlight. Necessarily, therefore, while the sunbeams make one side bright the other side is dark; and not only this, but the rings throw broad and heavy shadows upon Saturn in the direction away from the sunlight.

In the daytime they probably give out a faint shining, something like our own moon when seen in sunlight. During the summer nights they shine, no doubt, very beautifully. During the winter nights it so happens that their bright side is turned away; and not only that, but during the winter days the rings, while giving no light themselves to the wintry hemisphere of Saturn, completely hide the sun.

When it is remembered that Saturn's winter—that is, the winter of each hemisphere in turn—lasts during fifteen of our years; and when we hear of total eclipses of the sun lasting unbroken through eight years of such a winter, with not even bright rings to make up for his absence, we can not think of Saturn as a tempting residence. The sun gives Saturn at his best only about one-ninetieth of the heat and light that he gives to our earth; but to be deprived of even that little for eight years at a time, does indeed sound somewhat melancholy.

Looking now at the other side of the question, the possible inhabitants of the moons, especially those near at hand, would have splendid views of Saturn and his rings in all their varying phases. For Saturn is a beautiful globe, wrapped in his changeful envelope of clouds, which, seen through a telescope, are lit up often with rainbow tints of blue and gold; a creamy white belt lying usually on the equator; while around extend the purple and shining and gray rings, sometimes rivaling in bright colors Saturn himself.

We are compelled to assume that the continuous appearance of the rings is not due to real continuity of substance, but that they are composed of flights of disconnected satellites, so small and so closely packed that, at the immense distance to which Saturn is removed, they appear to form a continuous mass. There are analogous instances in the Solar System. In the zone of asteroids we have an undoubted instance of a flight of disconnected bodies traveling in a ring about a central attracting mass. The existence of zones of meteorites traveling around the sun has long been accepted as the only probable explanation of the periodic returns of meteoric showers. Again, the singular phenomenon called the Zodiacal Light is, in all probability, caused by a ring of minute cosmical bodies surrounding the sun. In the Milky Way and in the ring-nebulæ we have other illustrations of similar arrangements in nature, belonging, however, to orders immeasurably vaster than any within the Solar System.

The moons of Saturn do not, like those of Jupiter, travel in one plane.

CHAPTER XIX.

URANUS AND NEPTUNE.

TILL the year 1781, Saturn was believed to be the outermost planet of the Solar System, and nobody suspected the fact of two great, lonely brother-planets wandering around the same sun at vast distances beyond,—Uranus, nine hundred millions of miles from Saturn; Neptune, nine hundred millions of miles from Uranus. No wonder they remained long undiscovered.

Uranus can sometimes be seen by the unaided eye as a dim star of the sixth magnitude. And when he was known for a planet, it was found that he *had* been often so seen and noted. Again and again he had been taken for a fixed star, and as he moved on, disappearing from that particular spot, it was supposed that the star had vanished.

One night, March 13, 1781, when William Herschel was busily exploring the heavens with a powerful telescope, he noticed something which he took for a comet without a tail. He saw it was no mere point of light like the stars, but had a tiny round disk or face, which could be magnified. So he watched it carefully, and found in the course of a few nights that it moved—very slowly, certainly; but still it did move. Further watching and calculation made it clear that, though the newly-found heavenly body was at a very great distance from the sun, yet it was moving slowly in an orbit *round* the sun. Then it was known to be a planet, and another member of the Solar System.

From that day the reputation of Herschel rapidly increased. King George III, who loved the sciences and patronized them, had the astronomer presented to him; charmed with the simple and modest account of his efforts and his labors, he secured him a life pension and a residence at Slough, in the neighborhood of Windsor Castle. His sister Caroline assisted him as secretary, copied all his observations, and made all his calculations; the king gave him the title and salary of assistant astronomer. Before long the observatory at Slough surpassed in celebrity the principal observatories of Europe; we may say that, in the whole world, this is the place where the most discoveries have been made.

Astronomers soon applied themselves to the observation of the new body. They supposed that this "comet" would describe, as usually happens, a very elongated ellipse, and that it would approach considerably to the sun at its perihelion. But all the calculations made on this supposition had to be constantly recommenced. They could never succeed in representing all its positions, although the star moved very slowly: the observations of one month would utterly upset the calculations of the preceding month.

Several months elapsed without a suspicion that a veritable planet was under observation, and it was not till after recognizing that all the imaginary orbits for the supposed comet were contradicted by the observations, and that it had probably a circular orbit much farther from the sun than Saturn, till then the fron-

PROMINENT ASTRONOMERS OF FORMER TIMES.

tier of the system, that astronomers came to consider it as a planet. Still, this was at first but a provisional consent.

It was, in fact, more difficult than we may think to increase without scruple the family of the sun. Indeed, for reasons of expediency this idea was opposed. Ancient ideas are tyrannical. Men had so long been accustomed to consider old Saturn as the guardian of the frontiers, that it required a rare boldness of spirit to decide on extending these frontiers and marking them by a new world.

William Herschel proposed the name of *Georgium Sidus* "the Star of George," just as Galileo had given the name of "Medicean stars" to the satellites of Jupiter discovered by him, and as Horace had said *Julium Sidus*. Others proposed the name of *Neptune*, in order to maintain the mythological character, and to give to the new body the trident of the English maritime power; others, *Uranus*, the most ancient deity of all, and the father of Saturn, to whom reparation was due for so many centuries of neglect. Lalande proposed the name of *Herschel*, to immortalize the name of its discoverer. These last two denominations prevailed. For a long time the planet bore the name of Herschel; but custom has since declared for the mythological appellation, and Jupiter, Saturn, and Uranus succeed each other in order of descent—son, father, and grandfather.

The discovery of Uranus extended the radius of the Solar System from 885,000,000 to 1,765,000,000 of miles.

The apparent brightness of this planet is that of

a star of the sixth magnitude; observers whose sight is very piercing may succeed in recognizing it with the naked eye when they know where to look for it. Uranus moves slowly from west to east, and takes no less than eighty-four years to make the complete circuit of the sky. In its annual motion round the sun the earth passes between the sun and Uranus every 369 days—that is to say, once in a year and four days. It is at these times that this planet crosses the meridian at midnight. We can observe it in the evening sky for about six months of every year.

Everybody supposed that now, at least, the outermost member of all was discovered. But a very strange and remarkable thing happened.

Astronomers know with great exactness the paths of the planets in the heavens. They can tell, years beforehand, precisely what spot in space will be filled at any particular time by any particular planet. I am speaking now of their movements round the sun and in the Solar System—not of the movements of the whole family with the sun, about which little is yet known.

Each planet has its own particular pathway; its own particular distance from the sun, varying at each part of its pathway; its own particular speed in traveling round the sun, changing constantly from faster to slower or slower to faster, according to its distance from the sun, and according to the pull backwards or forwards of our neighboring planets in front or in rear. For as the orbits of all the planets are ovals, with the sun not in the middle, but somewhat to one side of the middle, it follows that all the planets, in the course of their years, are sometimes nearer to and sometimes farther from the sun.

The astronomers of the present day understand this well, and can describe with exactness the pathway of each planet. This knowledge does not come merely from watching one year how the planets travel, and remembering for another year, but is much more a matter of close and difficult calculation. Many things have to be considered, such as the planet's distance from the sun, the sun's power of attraction, the planet's speed, the nearness and weight of other neighboring planets.

All these questions were gone into, and astronomers sketched out the pathway in the heavens which they expected Uranus to follow. He would move in such and such an orbit, at such and such distances from the sun, and at such and such rates of speed.

But Uranus would not keep to these rules. He quite discomfited the astronomers. Sometimes he went fast, when, according to their notions, he ought to have gone more slowly; and sometimes he went slowly, when they would have looked for him to go more fast, and the line of his orbit was quite outside the line of the orbit which they had laid down. He was altogether a perplexing acquaintance, and difficult to understand. However, astronomers felt sure of their rules and modes of calculation, often before tested, and not found to fail. They made a guess at an explanation. What if there were yet another planet beyond Uranus, disturbing his motions—now drawing him on, now dragging him back, now so far balancing the sun's attraction by pulling in the opposite direction as to increase the distance of Uranus from the sun?

It might be so. But who could prove it? Hundreds of years might pass before any astronomer, in his star-gazing, should happen to light upon such a dim and distant world. Nay, the supposed planet might be, like Uranus, actually seen, and only be mistaken for a "variable star," shining but to disappear.

There the matter seemed likely to rest. There the matter probably would have rested for a good while, had not two men set themselves to conquer the difficulty. One was a young Englishman, a student of Cambridge, John Couch Adams; the other, a young Frenchman, Urbain Jean Joseph Leverrier,—both being astronomers.

Each worked independently of the other, neither knowing of the other's toil. The task which they had undertaken was no light one—that of reaching out into the unknown depths of space to find an unknown planet.

Each of these silent searchers into the sky-depths calculated what the orbit and speed of Uranus would be, without the presence of another disturbing planet beyond. Each examined what the amount of disturbance was, and considered the degree of attraction needful to produce that disturbance, together with the direction from which it had come. Each, in short, gradually worked his way through calculations far too deep and difficult for ordinary minds to grasp, till he had found just that spot in the heavens where a planet *ought* to be, to cause, according to known laws, just such an effect upon Uranus as had been observed.

Adams finished his calculation first, and sent the result to two different observatories. Unfortunately, his report was not eagerly taken up. It was, in fact, hardly believed. Leverrier finished his calculation also, and sent the result to the Berlin Observatory. The planet was actually seen in England first, but the discovery was actually made known from Berlin first. The young Englishman had been beforehand, but the young Frenchman gained foremost honor.

It was on the 23d of September, 1846, that Leverrier's letter reached the Berlin astronomers. The sky that night was clear, and we can easily understand with what anxiety Dr. Galle directed his telescope to the heavens. The instrument was pointed in accordance with Leverrier's instructions. The field of view showed, as does every part of the heavens, a multitude of stars. One of these was really the planet. The new chart—prepared by the Berlin Academy of Sciences, upon which the place of every star, down even to those of the tenth magnitude, was engraved—was unrolled, and, star by star, the heavens were compared with the chart. As the process of identification went on, one object after another was found in the heavens as engraved on the chart, and was of course rejected. At length a star of the eighth magnitude—a brilliant object—was brought into review. The map was examined, but there was no star there. This object could not have been in its present place when the map was formed. The object was therefore a wanderer—a planet. Yet it is necessary to be excessively cautious in such a matter.

Many possibilities had to be guarded against. It

was, for instance, possible that the object was really a star which, by some mischance, eluded the careful eye of the astronomer who constructed the map. It was even possible that the star might be one of the large class of variables which alternate in brightness, and it might have been conceivable that it was too faint to be seen when the chart was made. Even if neither of these explanations would answer, it was still necessary to show that the object was now moving, and moving with that particular velocity and in that particular direction which the theory of Leverrier indicated. The lapse of a single day was sufficient to dissipate all doubts. The next night the object was again observed. It had moved, and when its motion was measured it was found to accord precisely with what Leverrier had foretold. Indeed, as if no circumstance in the confirmation should be wanting, it was ascertained that the diameter of the planet, as measured by the micrometers at Berlin, was practically coincident with that anticipated by Leverrier.

The world speedily rang with the news of this splendid achievement. Instantly the name of Leverrier rose to a pinnacle hardly surpassed by that of any astronomer of any age or country. For a moment it seemed as if the French nation were to enjoy the undivided honor of this splendid triumph. But in the midst of the pæans of triumph with which the enthusiastic French nation hailed the discovery of Leverrier, there appeared a letter from Sir John Herschel in the *Athenæum* for 3d October, 1846, in which he announced the researches made by Adams, and claimed for him a participation in the glory of the discovery.

Subsequent inquiry has shown that this claim was a just one, and it is now universally admitted by all independent authorities.

This, however, was of slight comparative importance. The truly wonderful part of the matter was, that these two men could have so reasoned that, from the movements of one lately-discovered planet, they could point out the exact spot where a yet more distant planet ought to be, and that close to this very spot the planet was found. For when, both in England and in Germany, powerful telescopes were pointed in the direction named—*there the planet was.* No doubt about the matter. Not a star, but a real new planet in the far distance, wandering slowly round the sun.

Here we have the discovery of Neptune in its simple grandeur. This discovery is splendid, and of the highest order from a philosophical point of view, for it proves the security and the precision of the data of modern astronomy. Considered from the point of view of practical astronomy, it was but a simple exercise of calculation, and the most eminent astronomers saw in it nothing else! It was only after its verification, its public demonstration—it was only after the visual discovery of Neptune—that they had their eyes opened, and felt for a moment the dizziness of the infinite in view of the horizon revealed by the Neptunian perspective. The author of the calculation himself, the transcendent mathematician, did not even give himself the trouble to take a telescope and look at the sky to see whether a planet was really there! I even believe that he never saw it. For him, however, then and always, to the end of his life, astronomy was entirely inclosed in formulæ—the stars were but centers of force.

Very often I submitted to him the doubts of an anxious mind on the great problems of infinitude: I asked him if he thought that the other planets might be inhabited like ours, what might be especially the strange vital conditions of a world separated from the sun by the distance of Neptune, what might be the retinue of innumerable suns scattered in immensity, what astonishing colored lights the double stars should shed on the unknown planets which gravitate in these distant systems. His replies always showed me that these questions had no interest for him, and that, in his opinion, the essential knowledge of the universe consisted in equations, formulæ, and logarithmic series having for their object the mathematical theory of velocities and forces.

But it is not the less surprising that he had not the *curiosity* to verify the position of his planet, which would have been easy, since it shows a planetary disk; and, besides, he might have had the aid of a chart, because he had only to ask for these charts from the Berlin Observatory, where they had just been finished and *published*. It is not the less surprising that Arago, who was more of a physicist than a mathematician, more of a naturalist than a calculator, and whose mind had so remarkable a synthetical character, had not himself directed one of the telescopes of the observatory towards this point of the sky, and that no other French astronomer had this idea. But what surprises us still more is to know that *nearly a year be-*

fore, in October, 1845, a young student of the University of Cambridge, Mr. Adams, had sought the solution of the *same* problem, obtained the same results, and communicated these results to the director of the Greenwich Observatory, and that the astronomer to whom these results were confided had said nothing, and had not himself searched in the sky for the optical verification of his compatriot's solution!

This was indeed a triumph of human intellect. Yet it is no matter for human pride, but rather of thankfulness to God, who gave to man this marvelous reasoning power. And the very delight we have in such a success may humble us in the recollection of the vast amount lying beyond of the utterly unknown. But while the rare mind of a Newton could meekly realize the littleness of all he knew, lesser minds are very apt to be puffed up with the thought of what the human intellect can accomplish.

Perhaps the chief feeling of lawful satisfaction in this particular discovery arises from the fact that it gives marked and strong proof of the truth of our present astronomical system and beliefs. Many mistakes may be made, and much has often to be unlearned. Nevertheless, if the general principles of modern astronomy were wrong, if the commonly-received facts were a delusion, such complete success could not have attended so delicate and difficult a calculation.

We do not know much about these two outer planets, owing to their enormous distance from us. Uranus is in size equal to seventy-four earths, and Neptune is in size equal to one hundred and five earths. Both these planets are formed of somewhat heavier materials than Saturn, being about as dense as water.

The size of the sun as seen from Uranus is about one three-hundred-and-ninetieth part of the size of the sun we see. To Neptune he shows a disk only one nine-hundredth part of the size of that visible to us—no disk at all, in fact, but only starlike brilliancy.

The Uranian year lasts about eighty-four of our years; and this, with a very sloping axis, must cause most long and dreary winters, the tiny sun being hidden from parts of the planet during half an earthly life-time.

Uranus has at least four moons, traveling in very different planes from the plane of the ecliptic. Once it was thought that he had eight, but astronomers have since searched in vain for the other four, believed for a while to exist. Neptune has one moon, and may possess others not yet discovered.

Sir Henry Holland, the celebrated physician, and a devoted student of astronomy, has left on record an incident of his life connected with the planet Neptune of singular interest. I give it here, and in his own words, because it is scarcely likely otherwise to fall under the notice of astronomical readers. After stating that his interest in astronomy had led him to take advantage of all opportunities of visiting foreign observatories, he says:

"Some of these opportunities, indeed, arising out of my visits to observatories both in Europe and America, have been remarkable enough to warrant a more particular mention of them. That which most strongly clings to my memory is an evening I passed with Encke and Galle in the observatory at Berlin, some ten or twelve days after the discovery of the planet Neptune on this very spot; and when every night's observations of its motions had still an especial value in denoting the elements of its orbit. I had casually heard of the discovery at Bremen, and lost no time in hurrying on to Berlin. The night in question was one of floating clouds, gradually growing into cumuli; and hour after hour passed away without sight of the planet which had just come to our knowledge by so wonderful a method of predictive research. Frustrated in this main point, it was some compensation to stay and converse with Encke in his own observatory, one signalized by so many discoveries, the stillness and darkness of the place broken only by the solemn ticking of the astronomical clock, which, as the unfailing interpreter of the celestial times and motions, has a sort of living existence to the astronomer. Among other things discussed, while thus sitting together in a sort of tremulous impatience, was the name to be given to the new planet. Encke told me he had thought of 'Vulcan,' but deemed it right to remit the choice to Leverrier, then supposed to be the sole indicator of the planet and its place in the heavens; adding that he expected Leverrier's answer by the first post. Not an hour had elapsed before a knock at the door of the observatory announced the letter expected. Encke read it aloud; and, coming to the passage where Leverrier proposed the name of 'Neptune,' exclaimed, '*So lass den Namen Neptun sein.*'

It was a midnight scene not easily to be forgotten. A royal baptism, with its long array of titles, would ill compare with this simple naming of the remote and solitary planet thus wonderfully discovered. There is no place, indeed, where the grandeur and wild ambitions of the world are so thoroughly rebuked and dwarfed into littleness as in the astronomical observatory. As a practical illustration of this remark, I would add that my own knowledge of astronomers —those who have worked themselves with the telescope—has shown them to be generally men of tranquil temperament, and less disturbed than others by worldly affairs, or by the quarrels incident even to scientific research."

The same reasoning which has been used in reference to Jupiter and his moons, and to Saturn and his moons, might perhaps be applied also to Uranus and Neptune and their moons.

We know too little yet of their condition to venture far in such speculations. Still, taking the matter as a whole, there seem many reasons to incline us to the idea that each of the four greater outside planets may be a kind of secondary half-cooled sun to his satellites, helping to make up to them for the small amount of light and heat which they can obtain from the far-off sun.

CHAPTER XX.

COMETS AND METEORITES.

A CURIOUS discovery has been lately made. It is that some sort of mysterious tie seems to exist between comets and meteorites. For a long while this was never suspected. How should it be? The comets, so large, so airy, so light; the meteorites, so small, so solid, so heavy,—how could it possibly be supposed that the one had anything to do with the other? But supposings often have to give in to facts. Astronomers are gradually becoming convinced that there certainly is a connection between the two.

Strange to say, comets and meteorites occupy—sometimes, at least—the very same pathways in the heavens, the very same orbits round the sun. A certain number of meteorite systems are now pretty well known to astronomers as regularly met by our earth at certain points in her yearly journey. Some of these systems or rings have each a comet belonging to it—not merely journeying near, but actually in its midst, on the same orbit.

Perhaps it would be more correct to say that the meteorites belong to the comet, than that the comet belongs to the meteorites. We tread here on uncertain ground; for whether the meteorites spring from the comet, or whether the comet springs from the meteorites, or whether each has been brought into existence independently of the other, no one can at present say.

Though it is out of our power to explain the kind of connection, yet a connection there plainly is. So many instances are now known of a comet and a meteorite ring traveling together that it is doubtful whether any such ring could be found without a comet in its midst. By and by the doubt may spring up whether there ever exists a comet without a train of meteorites following him.

Among the many different meteorite rings which are known, two of the most important are the so-called August and November systems. Of these two, the November system must claim our chief attention. Not that we are at all sure of these being the most important meteorite rings in the Solar System. On the contrary, as regards the November ring, we have some reason to think that matters may lie just the other way.

The comet belonging to the November system is a small one—quite an insignificant little comet—only visible through a telescope. We do not, of course, know positively that larger comets and greater meteorite systems generally go together, but, to say the least, it seems likely. And if the greatness of a ring can at all be judged of by the size of its comet, then the November system must be a third-rate specimen of its kind. It is of particular importance to us, merely because it happens to be the one into which our earth plunges most deeply, and which we therefore see and know the best. The August ring is, on the contrary, connected with a magnificent comet, and may be a far grander system. But our pathway does not lead us into the midst of the August meteors as into those of November. We pass, seemingly, through its outskirts.

The meteorites of the November system are very small. They are believed to weigh commonly only a few grains each. If they were larger and heavier, some of them would fall to earth as aerolites, not more than half-burnt in their rush through the atmosphere. But this they are never found to do.

There are known meteorite systems, which our earth merely touches or grazes in passing, from which drop aerolites of a very different description—large, heavy, solid masses. It is well for us that we do not plunge into the midst of any such ring, or we might find our air, after all, a poor protection.

The last grand display of the November system of meteorites took place in the years 1866 to 1869, being continued more or less during three or four Novembers following. The next grand display will occur in the year 1899.

For this system—Leonides, as it is called, because the falling-stars in this display seem all to shoot towards us from a spot in the constellation Leo—seems to have a "time" or "year" of thirty-three and a quarter earthly years. The shape of its orbit is a very long ellipse, near one end of which is the sun, while the other end is believed to reach farther away than the orbit of Uranus.

A great deal of curiosity has been felt about the actual length and breadth and depth of the stream of meteorites through which our solid earth has so often plowed her way. During many hours at a time, lookers-on have watched the magnificent display of heavenly fireworks,—not a mere shooting-star here and there, as on common nights, but radiant meteors, flashing and dying by thousands through the sky. In 1866 no less than eight thousand meteors, in two hours and a quarter, were counted from the Greenwich Observatory in England. A natural wonder sprang up in many minds as to the extent of the ring from which they fell.

For not in one night only, but in several nights during three or four years, and that not once only but once in every thirty-three years, thousands and tens of thousands appear to have been stolen by our earth from the meteorite-ring, never again to be restored. Yet each time we touch the ring, we find the abundance of little meteorites in nowise seemingly lessened.

When speaking of a "ring" of meteorites, it must not be supposed that necessarily the meteorites form a whole unbroken ring all round the long oval orbit. There may be no breaks. There may be a more or less thin scattering throughout the entire length of the pathway. But the meteorites certainly seem to cluster far more densely in some parts of the orbit than in other parts, and it was about the size of the densest cluster that so much curiosity was felt.

Little can be positively known, though it is very certain that the cluster must be enormous in extent. Three or four years running, as our earth, after journeying the whole way round the sun, came again to that point in her orbit where she passes through the orbit of the Leonides, she found the thick stream of meteorites still pouring on, though each year lessening in amount. Taking into account this fact, and also the numbers that were seen to fall night after night, and also the speed of our earth, a "rough estimate" was formed.

The length of this dense cluster is supposed to reach to many hundreds of millions of miles. The thickness or depth of the stream is calculated to be in parts over five hundred thousand miles, and the breadth about ten times as much as the depth. Each meteorite is probably at a considerable distance from his neighbor; but the whole mass of them, when in the near neighborhood of the sun, must form a magnificent sight.

And if this be only a third-rate system, what must a first-rate system be like? And how many such systems are there throughout the sun's wide domains? The most powerful telescope gives us no hint of the existence of these rings till we find ourselves in their midst.

It may be that they are numbered by thousands, even by millions. The whole of the Solar System—nay, the very depths of space beyond—may, for aught we know, be crowded with meteorite systems. Every comet may have his stream of meteorites following him. But though the comet is visible to us, the meteorites are not. Billions upon billions of them may be ever rushing round our sun, entirely beyond our ken, till one or another straggler touches our atmosphere to flash and die as a "shooting star" in our sight.

We have, and with our present powers we can

have, no certainty as to all this. But I may quote here the illustration of a well-known astronomical writer on the subject.

Suppose a blind man were walking out of doors along a high road, and during the course of a few miles were to feel rain falling constantly upon him. Would it be reasonable on his part, if he concluded that a small shower of rain had accompanied him along the road as he moved, but that fine weather certainly existed on either side of the road? On the contrary, he might be sure that the drops which he felt, were but a few among millions falling all together.

Or, look at the raindrops on your window some dull day at home. Count how many there are? Could you, with any show of common sense, decide that those raindrops, and those alone, had fallen that day? So, when we find these showers of meteorites falling to earth, we may safely conclude that, for every one which touches our atmosphere, myriads rush elsewhere in space, never coming near us.

CHAPTER XXI.

MORE ABOUT COMETS AND METEORITES.

THE three prominent features of comets—a head, nucleus, and tail—are seen only in particular instances. The great majority appear as faint globular masses of vapor, with little or no central condensation, and without tails. On the other hand, some have a strongly-defined nucleus, which shines with a light as vivacious as that of the planets, so as to be even visible in the daytime. This was the case with one seen at Rome soon after the assassination of Julius Cæsar, believed by the populace to be the soul of the dictator translated to the skies. The first comet of the year 1442 was also so brilliant, that the light of the sun at noon, at the end of March, did not prevent it from being seen; and the second, which appeared in the summer, was visible for a considerable time before sunset. In 1532, the people of Milan were alarmed by the appearance of a star in the broad daylight; and as Venus was not then in a position to be visible, the object is inferred to have been a comet. Tycho Brahe discovered the comet of 1577, from his observatory in the isle of Huene, in the sound, before sunset; and Chizeaux saw the comet of 1744, at one o'clock in the afternoon, without a telescope.

The comet of 1843, already referred to in a previous chapter, was distinctly seen at noon by many persons in the streets of Bologna without the aid of glasses.

The nucleus is generally of small size, but the surrounding nebulosity, which forms the head, is often of immense extent; and the hazy envelope, the "horrid hair" of poetry, is sometimes seen separated from it by a dark space, encircling it like a ring. This was the aspect of the comet of 1811. The entire diameter of

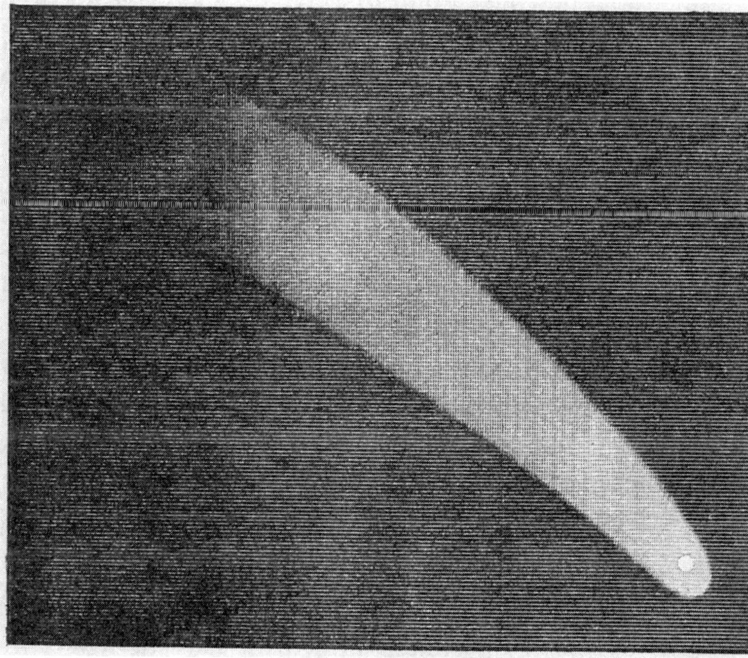

THE GREAT COMET OF 1811.

the head measured 1,250,000 miles, and hence its bulk was nearly three times that of the sun, and four million times that of the earth. But such extraordinary dimensions are quite exceptional.

Popular impressions respecting the supposed terrestrial effects of the comet of 1811—a fine object in the autumn of that year, long remembered by many—are on record. It was gravely noted, that wasps were very few in number; that flies became blind, and disappeared early; and that twins were born more frequently than usual. The season was remarkable for its bountiful harvest and abundant vintage. Grapes, figs, melons, and other fruits, were not only produced in extraordinary quantity, but of delicious flavor, so that "comet wines" had distinct bins allotted to them in the cellars of merchants, and were sold at high prices. There is, however, no fact better attested, by a comparison of observations, than that comets have no influence whatever in heightening or depressing the temperature of the seasons. The fine fruits and ample harvest of the year in question were, therefore, coincidences merely with the celestial phenomenon, without the slightest physical connection with it.

Though the advanced civilization of recent times has led to juster views of cometary apparitions than to regard them as divinely-appointed omens of terrestrial calamity, yet society is apt to be nervous respecting these bodies, as likely to cause some great natural convulsion by collision with our globe.

The great comet, which in 1682 received first the name of Halley's comet, appeared last in 1835, his return having been foretold within three days of its actually taking place. For Halley's comet is a member of the Solar System, having a yearly journey of seventy-six earthly years. He journeys nearer the sun than Venus, and travels farther away than Neptune.

In the years 1858, 1861, and 1862, three more comets appeared, all visible without the help of a telescope. Of these three, Donati's comet, in 1858, was far superior to the rest. This fine object was first discovered on the 2d of June by Dr. Donati, at Florence. It appeared at first as a faint nebulous patch of light, without any remarkable condensation; and as its motion towards the sun at that time was slow, it was not till the beginning of September that

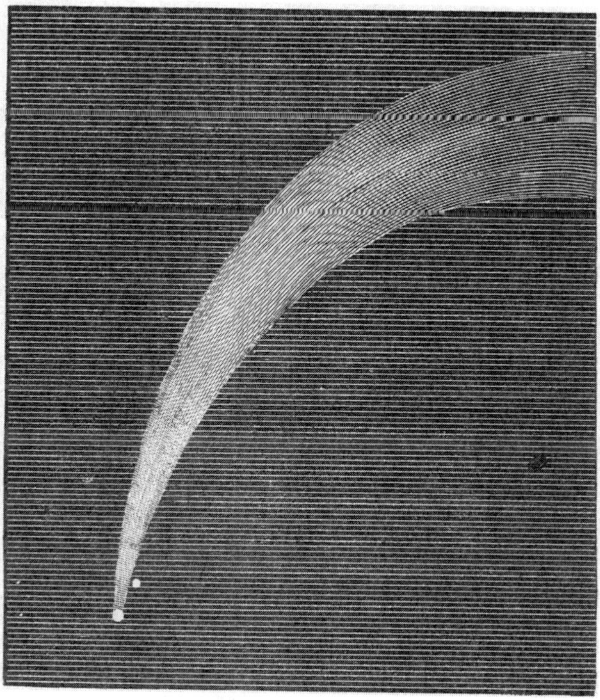

DONATI'S COMET, 1858.

it became visible to the naked eye. A short tail was then seen on the side opposite to the sun.

The development was afterwards rapid, and the comet became equally interesting to the professional astronomer and to the unlearned gazer. In the first week of October the tail attained its greatest length, 36°. At this time the tail was sensibly curved, and had the appearance of a large ostrich-feather when waved gently in the hand. At the latter end of October the comet was lost in the evening twilight. Its nearest approach to the earth was about equal to half the distance of the earth from the sun, and the nearest approach to Venus was about one-ninth part of that distance. An elliptical orbit, with a periodic time of more than 2,000 years, has been assigned to our late visitor.

It was a singular fact about this comet that at one time the star Arcturus could be seen shining through the densest portion of the tail, close to the nucleus. Now, although the faintest cloud-wreath of earth would dim, if not hide, this star, yet the tail of the great comet was of so transparent a nature that Arcturus shone undimmed, as if no veil had come between. The exceedingly slight and airy texture of a comet's tail could hardly be more plainly shown. It was this gauzy appendage to a little nucleus which men once thought could destroy our solid earth at a single blow!

These cometary trains generally appear straight, or at least, by the effect of perspective, they seem to be directed along the arcs of great circles of the celestial sphere. Some are recorded, however, which presented a different appearance. Thus, in 1689, a comet was seen whose tail, according to the historians, was curved like a Turkish saber. This tail had a total length of 68°. It was the same with the beautiful comet of Donati, which we admired in 1858, and of which the tail had a very decided curvature.

A comet can only be observed in the sky for a

limited time. We perceive it at first in a region where nothing was visible on preceding days. The next day, the third day, we see it again; but it has changed its place considerably among the constellations. We can thus follow it in the sky for a certain number of days, often during several months; then we cease to perceive it. Often the comet is lost to view because it approaches the sun, and the vivid light of that body hides it completely; but soon we observe it again on the other side of the sun, and it is not till some time afterwards that it definitely disappears. Some have been visible at noonday, quite near the sun, like that of 1882, on September 17th.

Yet, while taking care not to overrate, we must not underrate. True, the comets are delicate and slight in structure. One comet, in 1770, wandered into the very midst of Jupiter's moons, and so small was its weight that it had no power whatever, so far as has been detected, to disturb the said moons in their orbits. Jupiter and his moons did very seriously disturb the comet, however; and when he came out from their midst, though none the worse for his adventure, he was forced to travel in an entirely new orbit, and never managed to get back to his old pathway again.

But there are comets *and* comets, some being heavier than others. The comet named after Donati, albeit too transparent to hide a star, was yet so immense in size that his weight was calculated by one astronomer to amount to as much as a mass of water forty thousand miles square and one hundred and nine yards deep.

When first noticed, Donati's comet had, like all large comets, a bright envelope of light round the nucleus. After a while the one envelope grew into three envelopes, and a new tail formed beside the principal tail, which for a time was seen to bend gracefully into a curve, like a splendid plume. A third, but much fainter tail, also made its appearance, and many angry-looking jets were poured out from the nucleus. These changes took place while the comet was passing through the great heat of near neighborhood to the sun. Afterwards, as he passed away, he seemed gradually to cool down and grow quiet. The singular changes in the appearance of Newton's comet have been earlier noticed. No marvel that he did undergo some alterations. The tremendous glare and burning heat which that comet had to endure in his rush past the sun, were more than twenty-five thousand times as much as the glare and heat of the fiercest tropical noonday ever known upon earth. Can we wonder that he should have shown "signs of great excitement," that his head should have grown larger and his tail longer?

It certainly was amazing, and past comprehension, that the said tail, over ninety millions of miles in length, should in four days have seemingly swept round in a tremendous half-circle, so as first to point in one direction, and then to point in just the opposite direction.

We are much in the habit of speaking about comets as traveling through the heavens with their tails streaming behind them. But though this is sometimes the case, it is by no means always.

The tails of comets always stream *away from the sun*—whether before or behind the comet's head seeming to be a matter of indifference. As the comet comes hurrying along his orbit, with ever-increasing speed, towards the sun, the head journeys first, and the long tail follows after. But as the comet rounds the loop of his orbit near the sun—the point nearest of all being called his *perihelion*—the head always remains towards the sun, while the tail swings, or seems to swing, in a magnificent sweep round, pointing always in the direction just away from the sun.

Then, as the comet journeys with slackening speed, on the other side of his orbit, towards the distant *aphelion*, or farthest point from the sun, he still keeps his head towards the sun. So, at this part of his passage, in place of the head going first and the tail following after, the tail goes first and the head follows after. The comet thus appears to be moving backwards; or, like an engine pushing instead of drawing a train, the head seems to be driving the tail before it.

The sun, then, acts on these bodies when they approach him, produces in them important physical and chemical transformations, and exercises on their developed atmosphere a repulsive force, the nature of which is still unknown to us, but the effects of which coincide with the formation and development of the tails. The tails are thus, in the extension of the cometary atmosphere, driven back, either by the solar heat, by the light, by electricity, or by other forces; and this extension is rather a motion in the ether than a real transport of matter, at least in the great comets which approach very near the sun, and in their immense luminous appendages. The effects produced and observed are not the same in all comets, which

GREAT COMET OF 1744.

proves that they differ from each other in several respects. The tails have sometimes been seen to diminish before the perihelion passage, as in 1835.

Luminous envelopes have also been seen succeeding each other round the head, concentrating themselves on the side opposite to the sun, and leaving the central line of the tail darker than the two sides. This is what happened in the Donati comet and in that of 1861. Sometimes a secondary tail has been seen projected towards the sun, as in 1824, 1850, 1851, and 1880. Comets have been seen with the head envel-

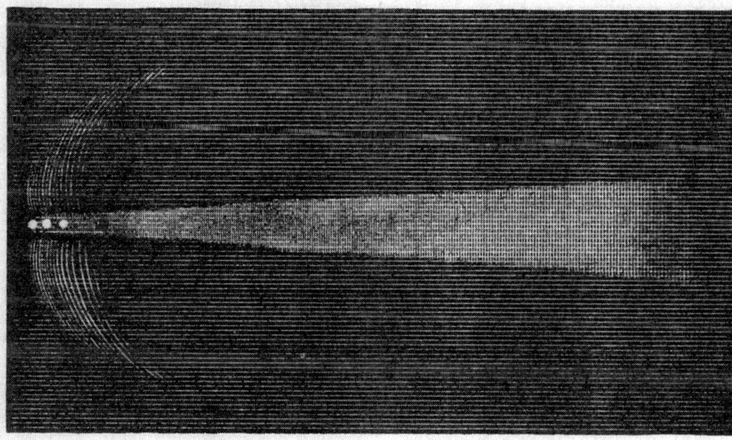

THE FIRST COMET OF 1888—JUNE 4.

oped in phosphorescence, surrounding them with a sort of luminous atmosphere. Comets have also been seen with three, four, five, and *six tails*, like that of 1744, for example, which appeared like a splendid aurora borealis rising majestically in the sky, until, the celestial fan being raised to its full height, it was perceived that the six jets of light all proceeded from the same point, which was nothing else but the nucleus of a comet. On the other hand, the nuclei themselves show great variations—some appear simply nebulous, and permit the faintest stars to be visible through them; others seem to be formed of one or more solid masses surrounded by an enormous atmosphere; in others, again, a nucleus does not exist, as in the Southern comet of 1887. One of the comets of 1888 showed a triple nucleus and a bristling coma, as may be seen in the cut. We may, then, consider that the wandering bodies collected under the name of comets are of several origins and *several different species*.

Why the tails of comets should so persistently avoid the sun it is impossible to say. We do not even know whether the cause lies actually in the sun or in the comet. Astronomers speak of the "repulsive energy" with which the sun "sweeps away" from his neighborhood the light vapory matter of which the tails are made. But the how and the wherefore of this strange seeming repulsion they can not explain. For at the selfsame time the sun appears to be attracting the comet towards himself, and driving the comet's tail away from himself.

A few of the more well-known comets have been mentioned; but a year rarely passes in which at least one comet is not discovered, though often only a small specimen, sometimes even tailless and hairless. No doubt many others pass unseen. The large and grand ones only come to view now and then.

What are comets and meteorites made of? Respecting comets, a good many ideas are put forth, and a good many guesses are made, as to "burning gas," "luminous vapor," "beams of light," and so on. But in truth, little is known about the matter. Respecting meteorites, we can speak more certainly. A good many meteorites have fallen to earth as aerolites, and have been carefully examined. They are found to contain nickel, cobalt, iron, phosphorus, and sometimes at least, a large supply of hydrogen gas.

Not only do solid aerolites fall half-burnt to the ground, but even when the meteorites are quite consumed in the air, the fine dust remaining still sinks earthward. This fine dust has been found upon mountain-tops, and has been proved by close examination to be precisely the same as the material of the solid aerolites.

A luminous body of sensible dimensions rapidly traverses space, diffusing on all sides a vivid light—like a globe of fire of which the apparent size is often comparable with that of the moon. This body usually leaves behind it a perceptible luminous train. On or immediately after its appearance it often produces an explosion, and sometimes even several successive explosions, which are heard at great distances. These explosions are also often accompanied by the division of the globe of fire into luminous fragments, more or less numerous, which seem to be projected in different directions. This phenomenon constitutes what is called a *meteor*, properly speaking, or a *bolide*. It is produced during the day as well as the night. The light which it causes in the former case is greatly enfeebled by the presence of the solar light, and it is only when it is developed with sufficient intensity that it can be perceived.

In the British Museum there is a superb collection of meteorites. They have been brought together from all parts of the earth, and vary in size from bodies not much larger than a pin's-head up to vast masses weighing many hundred pounds. There are also many models of celebrated meteorites, of which the originals are dispersed through various other museums.

Many of these objects have nothing very remarkable in their external appearance. If they were met with on the sea-beach, they would be passed by without more notice than would be given to any other stone. Yet what a history such a stone might tell us if we could only manage to obtain it! It fell; it was seen to fall from the sky; but what was its course anterior to that movement? Where was it one hundred years ago, one thousand years ago? Through what regions of space has it wandered? Why did it never fall before? Why has it actually now fallen? Such are some of the questions which crowd upon us as we ponder over these most interesting bodies. Some of these objects are composed of very characteristic materials. Take, for example, one of the more recent meteorites, known as the Rowton siderite. This body differs very much from the more ordinary kind of stony meteorite. It is an object which even a casual passer-by would hardly pass without notice. Its great weight would also attract attention, while if it be scratched or rubbed with a file, it would be found that it was not in any sense a stone, but that it was a mass of nearly pure iron. We know the circumstances under which that piece of iron fell to the earth. It was on the 20th of April, 1876, about 3.40 P. M., that a strange rumbling noise, followed by a startling explosion, was heard over an area of several miles in extent among the villages in Shropshire. About an hour after this occurrence, a farmer noticed the ground in one of his grass-fields to have been disturbed, and he probed the hole which the meteorite had made, and found it, still warm, about eighteen inches below the surface. Some men working at no great distance had actually heard the noise of its descent, but without being able to indicate the exact locality. This remarkable object weighs 7¾ pounds. It is an irregular angular mass of iron, though all its edges seem to have been rounded by fusion in its transit through the air, and it is covered with a thick black film of the magnetic oxide of iron, except at the point where it first struck the ground.

This siderite is specially interesting on account of its distinctly metallic character. Falls of the siderites, as they are called, are not so common as those of the stony meteorites; in fact, there are only a few known instances of meteoric irons having been actually seen to fall, while the falls of stony meteorites are to be counted in scores or in hundreds. The inference is that the iron meteorites are much less frequent than the stony ones.

It is a wonderful thought that we should really have these visitants from the sky, solid metal or showers of dust, coming to us from distant space.

If the dust of thousands of meteorites is always thus falling earthward, one would imagine that it must in time add something to the weight of the earth. And this actually is the case. During the last three thousand years, no less than one million tons of meteorite-dust must, according to calculation, have fallen to earth out of the sky. A million tons is of

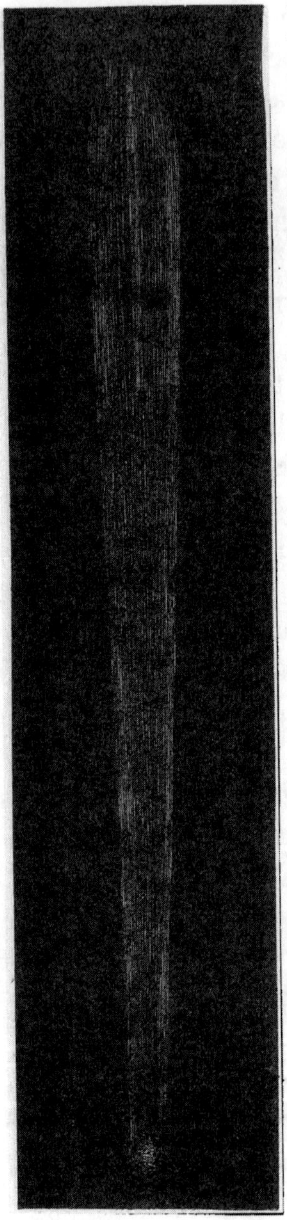

GREAT COMET OF 1680.

course a mere nothing compared with the size of the world. Still, the fact is curious and interesting.

It has been suggested that *perhaps* the flames of the sun are partly fed by vast showers of falling meteorites. It has even been suggested that *perhaps*, in long past ages, the earth and the planets grew to their present size under a tremendous downpour of meteorites; the numbers which now drop to earth being merely the thin remains of what once existed. But for this guess there is no real foundation.

Whether suns and worlds full-grown, created as in an instant; whether tiny meteorites in countless myriads to be used as stones in "building;" whether vast masses of flaming gas, to be gradually cooled and "framed" into shape—which ever may have come first, and whichever may have been the order of God's working—still that "first" was made by him; still he throughout was the Master-builder.

A few words more about "comet-visitors." Many comets, as already stated, belong to our

Solar System, though whether they have always so belonged is another question. It is not impossible that they may once upon a time have wandered hence from a vast distance, and, being caught prisoner by the powerful attraction of Jupiter or one of his three great brother-planets, have been compelled thenceforth to travel in a closed pathway round the sun.

There are also many comets which come once only to our system, flashing round past the sun, and rushing away in quite another direction, never to return. Where do these comets come from? And where do they go? From other suns—brother suns to ours? It may be. One is almost disposed to think that it must be so.

GREAT COMET OF 1769.

Do we ever think what an immense voyage

they must have made to come from there to here? Do we imagine for how many years they must have flown through the dark immensity to plunge themselves into the fires of our sun? If we take into account the directions from which certain comets come to us, and if we assign to the stars situated in that region the least distances consistent with known facts, we find that these comets certainly left their last star more than *twenty millions of years ago.*

In thus putting to us from the height of their celestial apparitions so many notes of interrogation on the grandest problems of creation, comets assume to our eyes an interest incomparably greater than that with which superstition blindly surrounded them in past ages. When we reflect for a moment that a certain comet which shines before us in the sky came originally from the depths of the heavens, that it has traveled during millions of years to arrive here, and that consequently it is by millions of years that we must reckon its age if we wish to form any idea of it, we can not refrain from respecting this strange visitor as a witness of vanished eras, as an echo of the past, as the most ancient testimony which we have of the existence of matter. But what do we say? These bodies are neither old nor young. There is nothing old, nothing new—all is present. The ages of the past contemplate the ages of the future, which all work, all gravitate, all circulate, in the eternal plan. Musing, you look at the river which flows so gently at your feet, and you believe you see again the river of your childhood; but the water of to-day is not that of yesterday, it is not the same substance which you have before your eyes, and never, never shall this union of molecules, which you behold at this moment, come back there—never till the consummation of the ages!

If the appearance of comets forebodes absolutely nothing as to the microscopical events of our ephemeral human history, it is not the same with the effects which might be produced by their encounter with our wandering planet. In such an encounter there is nothing impossible. No law of celestial mechanics forbids that two bodies should come into collision in their course, be broken up, pulverized, and mutually reduced to vapor.

I have before described some of the tremendous outbursts seen on the surface of our sun. It is believed that in these outbursts matter is expelled, or driven forth, with such fearful violence as to send it whirling through space, never to fall back to the sun.

Some meteorites may have their births thus. So also may some comets. And if all the stars are suns —huge fiery globes like our sun, and subject like him to tremendous eruptions—they too, probably, send out comets and meteorites to wander through space. Whether or no comets come into existence in any such manner, one thing seems pretty certain. Those comets which come to us from outside our system must come from some other system. And the nearest systems known are those of the stars.

The nearest star of all, whose distance has been measured, is Alpha Centauri. It has been roughly calculated that a comet, passing direct from Alpha Centauri to our sun, would take about twenty millions of years for his journey. But here we tread upon very doubtful ground. Many matters, at present unknown to us, might greatly affect the result of such a calculation. Also it is by no means impossible that

BARON ALEXANDER VON HUMBOLDT.

other stars lie really much nearer to us than Alpha Centauri, whose distance astronomers have not yet attempted to measure.

"In order to complete our view," says Alexander

Von Humboldt, "of all that we have learned to consider as appertaining to our Solar System, which now, since the discovery of the small planets, of the interior comets of short revolutions, and of the meteoric asteroids, is so rich and complicated in its form, it remains for us to speak of the ring of zodiacal light. Those who have lived for many years in the zone of palms must retain a pleasing impression of the mild radiance with which the zodiacal light, shooting pyramidally upward, illumines a part of the uniform length of tropical nights. I have seen it shine with an intensity of light equal to the Milky Way in Saggitarius, and that not only in the rare and dry atmosphere of the summit of the Andes, at an elevation of from thirteen to fifteen thousand feet, but even on the boundless grassy plains of Venezuela, and on the seashore, beneath the ever-clear sky of Cumana. This phenomenon was often rendered especially beautiful by the passage of light, fleecy clouds, which stood out in picturesque and bold relief from the luminous background.

"This phenomenon, whose primordial antiquity can scarcely be doubted, is not the luminous solar atmosphere itself, which can not be diffused beyond nine-tenths of the distance of Mercury. With much probability we may regard the existence of a very compressed ring of nebulous matter, revolving freely in space around the sun between the orbits of Venus and Mars, as the material cause of the zodiacal light. These nebulous particles may either be self-luminous or receive their light from the sun.

"I have occasionally been astonished, in the tropical climates of South America, to observe the variable intensity of the zodiacal light. As I passed the nights, during many months, in the open air, on the shores of rivers and on plains, I enjoyed ample opportunities of carefully examining this phenomenon. When the zodiacal light had been most intense, I have observed that it would be weakened for a few minutes, until it again suddenly shone forth in full brilliancy. In a few instances I have thought I could perceive, not exactly a reddish coloration, nor the lower portion darkened in an arc-like form, nor even a scintillation, but a kind of flickering and wavering of the light. Must we suppose that changes are actually in progress in the nebulous ring? Or is it not more probable that processes of condensation may be going on in the uppermost strata of the air, by means of which the transparency—or, rather, the reflection of light—may be modified in some peculiar and unknown manner? An assumption of the existence of such meteorological causes on the confines of our atmosphere is strengthened by the sudden flash and pulsations of light which have been observed to vibrate for several seconds through the tail of a comet. During the continuation of these pulsations it has been noticed that the comet's tail was lengthened by several degrees, and then again contracted."

CHAPTER XXII.

MANY SUNS.

ONCE more we have to wing our flight far, far away from the busy Solar System where we live; away from whirling planets, moons, meteorites, all shining with reflected sunlight; away from the great central sun himself, our own particular bright star. Once more we have, in imagination, to cross the vast, black, empty space—*is* it black, and *is* it empty, had we sight to see things as they are?—separating our sun from other suns, our star from other stars. For the sun is a star—only a star. And stars are suns—big, blazing suns. One is near, and the others are far away. That is the difference.

We have no longer to do with bodies merely reflecting another's light—always dark on one side and bright on the other—but with burning bodies, shining all round by their own light. We have no longer to picture just one single star with his surrounding worlds, but we have to fix our thoughts upon the great universe of stars or suns in countless millions.

The earth is forgotten, with its small and ephemeral history. The sun himself, with all his immense system, has sunk in the infinite night. On the wings of inter-sidereal comets we have taken our flight towards the stars, the suns of space. Have we exactly measured, have we worthily realized the road passed over by our thoughts? The nearest star to us reigns at a distance of 275,000 times our distance from the sun; out to that star an immense desert surrounds us, the most profound, the darkest, and the most silent of solitudes.

The solar system seems to us very vast; the abyss which separates our world from Mars, Jupiter, Saturn, and Neptune, appears to us immense; relatively to the fixed stars, however, our whole system represents but an isolated family immediately surrounding us: a sphere as vast as the whole solar system would be reduced to the size of a simple point if it were transported to the distance of the nearest star. The space which extends between the solar system and the stars, and which separates the stars from each other, appears to be entirely void of visible matter, with the exception of nebulous fragments, cometary or meteoric, which circulate here and there in the immense voids. Nine thousand two hundred and fifty systems like ours, bounded by Neptune, would be contained in the space which isolates us from the nearest star!

It is marvelous that we can perceive the stars at such a distance. What an admirable transparency in these immense spaces to permit the light to pass, without being wasted, to thousands of billions of miles! Around us, in the thick air which envelops us, the mountains are already darkened and difficult to see at seventy miles; the least fog hides from us objects on the horizon. What must be the tenuity, the rarefaction, the extreme transparency of the ethereal medium which fills the celestial spaces!

The sun is center and ruler and king in his own system. But as a star, he is only one among many stars, some greater, some less than himself. From the far siderial heavens he is hardly recognized even as a star among the constellations, so faint is his light.

Do the suns which surround us form a system with that which illuminates us, as the planets form one round our solar focus, and does our sun revolve round an attractive center? Does this center, the point of the revolutions of many suns, itself revolve round a preponderating center? In a word, is the visible universe organized in one or several systems? No Divine revelation comes to instruct men on the mysteries which interest them most—their personal or collective destinies. We have now, as ever, but science and observation to answer us.

A problem so vast as this is still far from receiving even an approximate solution. From whatever point of view we consider it, we find ourselves face to face with the infinite in space and time. The present aspect of the universe immediately brings into question its past and its future state, and then the whole of united human learning supplies us in this great research with but a pale light, scarcely illuminating the first steps of the dark and unknown road on which we are traveling. However, such a problem is worthy of engaging our attention, and positive science has already made sufficient discoveries in the knowledge of the laws of nature to permit us to attempt to penetrate these great mysteries. What is it that the general observation of the heavens—what is it that sidereal synthesis teaches us on our real situation in infinitude?

In the calm and silent hours of beautiful evenings, what pensive gaze is not lost in the vague windings of the Milky Way, in the soft and celestial gleam of that cloudy arch, which seems supported on two opposite points of the horizon, and elevated more or less in the sky according to the place of the observer and the hour of the night? While one-half appears above

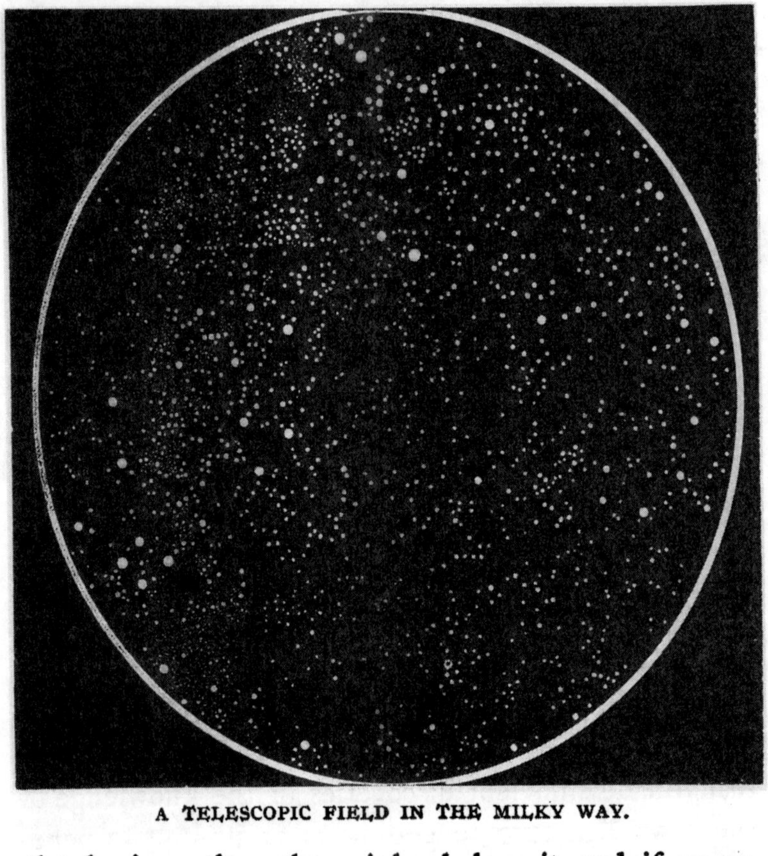

A TELESCOPIC FIELD IN THE MILKY WAY.

the horizon, the other sinks below it, and if we removed the earth, or if it were rendered transparent, we should see the complete Milky Way, under the form of a great circle, making the whole circuit of the sky. The scientific study of this trail of light, and its comparison with the starry population of the heavens, begins for us the solution of the great problem.

Let us point a telescope towards any point of this stupendous arch. Suddenly hundreds, thousands of stars show themselves in the telescopic field, like needle-points on the celestial vault. Let us wait for some moments, that our eye may become accustomed to the darkness of the background, and the little sparks shine out by thousands. Let us leave the instrument pointed motionless towards the same region, and there slowly passes before our dazzled vision the distant army of stars. In a quarter of an hour we see them appear by thousands and thousands. William Herschel counted three hundred and thirty-one thousand in a width of 5° in the constellation Cygnus, so nebulous to the naked eye. If we could see the whole of the Milky Way pass before us, we should see eighteen millions of stars.

This seed-plot of stars is formed of objects individually invisible to the naked eye, below the sixth magnitude; but so crowded that they appear to touch each other, and form a nebulous gleam which all human eyes, directed to the sky for thousands of years, have contemplated and admired. Since it is developed like a girdle round the whole circuit of the sky, we ourselves must be in the Milky Way. The first fact which impresses our mind is, that *our sun is a star of the Milky Way.*

Thought travels fast—faster than a comet, faster than light. A rushing comet would, it is believed, take twenty millions of years to cross the chasm between the nearest known fixed star and us. Light, flashing along at the rate of about one hundred and eighty-six thousand miles a second, will perform the same journey in four years and a third. But thought can overleap the boundary in less than a single moment.

Each star that we see in the heavens is to our eyesight simply one point of light. The brighter stars are said to be of greater magnitude, and the fainter stars of lesser magnitude; yet, one and all, they have no apparent size. The most powerful telescope, though it can increase their brilliancy, can not add to their size. A planet, which to the naked eye may look like a star, will, under a telescope, show a disk, the breadth of which can be measured or divided; but no star has any real, visible disk in the most powerful telescope yet constructed. The reason of this is the enormous distance of the stars. Far off as many of the planets lie, yet the farthest of them is as a member of our household compared with the nearest star.

I have already tried to make clear the fact of their vast distance. Light, which comes to us from the sun in eight minutes and a half, takes over four and a third years to reach us from Alpha Centauri. From this four-years-and-a-third length of journey between Alpha Centauri and earth, the numbers rise rapidly to twenty years, fifty years, seventy years, even hundreds of years. The distance of most of the stars is completely beyond our power to measure. The whole orbit of our earth, nay, the whole wide orbit of the far-off Neptune, would dwindle down to one single point, if seen from the greater number of the stars.

It used to be believed that, taking the stars generally, there was probably no very marked difference in their size, their kind, their brightness. Some of course would be rather larger, and others rather smaller; still it was supposed that they might be roughly classed as formed much on the same scale and the same plan. But doubts are now felt about this notion. For a very similar idea used to be held with regard to the Solar System. The wonderful variety of form and richness in numbers, now known to abound within its limits, are discoveries of late years. May not the same variety in kind and size be found also among the stars? The more we look into the heavens, the more we find that dull, blank uniformity is not to be seen there.

It is the same upon earth. Man builds his little rows of boxlike houses side by side, each one exactly like all the rest, or dresses his thousand soldiers in coats of the same cut and color, or repeats a neat leaf-design hundreds of times on carpets or wall-papers; but God never makes two leafs or two blades of grass alike. Wholesale turning out of things after one pattern is quite a human idea, not divine.

We know so much about the stars as that some are at least considerably larger and some considerably smaller than others. When one star is seen to shine brightly, and another beside it shines dimly, we are apt to think that the brightest must be the nearest. Yet it is often impossible for us to say how much of the difference is owing to the greater distance of one or the other, to the greater size of one or the other, or to the greater brilliancy of one or the other. In many instances we do know enough to be quite sure that there is a great difference, not only in the distance of the stars, but in their size, their kind, their brightness.

The stars have been lately classed by one or two astronomers into four distinct orders or degrees—partly depending on their color. The first class is that of the White Suns. These are said to be the grandest and mightiest of all. The star Sirius belongs to the order of White Suns. Secondly comes the class of Golden Suns. To these blazing furnaces of yellow light, second only to the white-light stars, belongs our own sun. Thirdly, there are numbers of stars called Variable Stars, the light of which is constantly changing, now becoming more, now becoming less. Fourthly, there is the class of small Red Suns, about which not much is known.

These four orders or divisions do not by any means include all the stars, or even all the single stars. Roughly speaking, however, the greater number of single stars, and many also of the double stars, belong to one or another of the above classes.

When we talk of the different sizes of the different stars, it should be plainly understood that we have no means of directly measuring them. A point of light showing no disk, no surface, no breadth, can not be measured, for there is nothing to measure.

In certain cases we are not entirely without the power of judging. The distances of a few of the stars from us have been found out. Knowing how far off any particular star is, astronomers are able to calculate exactly how bright our own sun would look at that same distance. If they find that our sun would shine just as the star in question shines, there is some reason for supposing that our sun and the star may be of the same size. If our sun would shine more brightly than the star shines, there is some reason for supposing that the star may be smaller than our sun. If our sun would shine more dimly than the star shines, there is some reason for supposing that the star may be larger than our sun,

Other matters, however, have to be considered. Suppose we find a star at a certain distance shining twice as brilliantly as our own sun would shine at that same distance. Naturally, then, we say, That star must be much larger than our sun.

The reasoning may be mistaken. We do not know the fact. What if, instead of being a much *larger* sun, it is only a much *brighter* sun? This possibly must be allowed for. It has, indeed, been strongly doubted by one astronomer, after close study of the sun, whether any surface of any star *could* exceed the surface of our sun in its power of light and heat.

But Sirius sheds actually forty-eight times as much light around it as does our sun; we are not exactly entitled to say that Sirius is forty-eight times as big as the sun, but we can say that Sirius is forty-eight times as brilliant or as splendid. In making this calculation we have taken a lower determination of the brightness of Sirius relatively to the sun than some other careful observations would have warranted. It will thus be seen that if there be any uncertainty in our result it must only be as to whether Sirius is not really more than forty-eight times as bright as the sun.

When we picture to ourselves the star-depths, the boundless reaches of heavenly space, with these countless blazing suns scattered broadcast throughout, we have not to picture an universe in repose. On the contrary, all is life, stir, energy. Just as in the busy whirl of our Solar System no such thing as rest is to be found, so also it seems to be in the wide universe.

Every star is in motion. "Fixed," as we call them, they are not fixed. Invisible as their movements are to our eyes through immensity of distance, yet all are moving. Those silent, placid, twinkling specks of light are, in reality, huge, roaring, seething, tumultuous furnaces of fire and flame, heat and radiance.

Each, too, is hurrying along his appointed pathway in space. Some move faster, some move more slowly. One mile per second; ten miles per second; twenty, thirty, forty, fifty miles per second,—thus varying are their rates of speed. But whether fast or whether slowly, still onward and ever onward they press. Some are rushing towards us, some are rushing away from us. Some are speeding to the right, some are speeding to the left.

The fine star Arcturus, which any one may admire every evening on the prolongation of the tail of the Great Bear, is slowly withdrawing from the fixed point where the celestial charts placed it two thousand years ago, and is moving towards the southwest. It takes eight hundred years to describe a space in the sky equal to the apparent diameter of the moon; nevertheless, this displacement was sufficiently perceptible to attract attention more than a century and a half ago, for Halley had noticed it in 1718, as well as those of Sirius and Aldebaran. However slow it may appear at the distance we are from Arcturus, this motion is, at a minimum, 1,637 millions of miles a year. Sirius takes 1,338 years to pass over the same angular space in the sky; at the distance of that star this is, at a minimum, 397 millions of miles per annum. The study of the proper motions of the stars has made the greatest progress in the last half-century, and especially in recent years. All the stars visible to the naked eye and a large number of telescopic stars show displacements of this kind.

Where are they going? Does any single star ever return to his starting-point—wherever that starting-point may have been? Do they journey in vast circles or ellipses, round some far-distant center? What controls them all? Is it the mighty power of some such center, or does each star, by his faint and distant attraction, help to control all his brother-stars, to guide them on their appointed path, to preserve the delicate balance of a universe?

How little we know about the matter! Only so much we can tell—that the controlling and restraining hand of God is over the whole. Whether by the attraction of one great center or by the united influences of a thousand fainter attractions, he steers each radiant sun upon its heavenly path, "upholding all things by the words of his power." There is no blundering, no confusion, no entanglement. All is perfect order, calm arrangement, restrained energy.

CHAPTER XXIII.

SOME PARTICULAR SUNS.

In the constellation of the Swan there is a little, dim, sixth-magnitude star, scarcely to be seen without a telescope. This star, 61 Cygni by name, is the first whose distance from us it was found possible to determine.

We may think it strange that so faint a star was even attempted. Would not astronomers have naturally supposed it to be one of the farther-distant stars? No, they did not. For though 61 Cygni showed but a dim light, yet his motion—not the daily apparent motion, but the real motion as seen from earth—was found to be so much more rapid than the motion of most other stars that they rightly guessed 61 Cygni to be a rather near neighbor of ours.

Do not misunderstand me, when I speak of "more rapid motion," and of "rather near neighborhood." The real rate of 61 Cygni's rush through space is believed to be about forty miles each second, or one thousand four hundred and fifty millions of miles each year. All we can perceive of this quick motion is that, in the course of three hundred and fifty years, 61 Cygni travels over a space in the sky about as long as the breadth of the full moon. Little enough, yet very far beyond what can be detected in the greater number of even the brightest stars.

Then, again, as to the near neighborhood of this

star, 61 Cygni is near enough to have his distance measured, and that is saying a good deal. Alpha Centauri, the nearest star of which we know in the southern heavens, is two hundred and seventy-five thousand times as far distant as the sun. But 61 Cygni, the nearest star of which we know in the northern heavens, is nearly twice as far away as Alpha Centauri.

We call 61 Cygni a star, for so he appears to common observers. In reality, instead of being only one star, the speck of light which we call 61 Cygni consists of *two stars*. The two are separated by a gap about half as wide again as the wide gap between the sun and Neptune. Yet so great is their distance from us that to the naked eye the two seem to be one. These two suns would together make a sun only about one-third as large as our sun. They differ in size, the quicker movements of one showing it to be the smaller; and it is by means of their known distance from one another, and their known rate of motion, that their size, or rather their weight, can be roughly calculated. Of course neither of the two shows any actual measurable disk. So much and so little is with tolerable certainty known about this particular pair of suns.

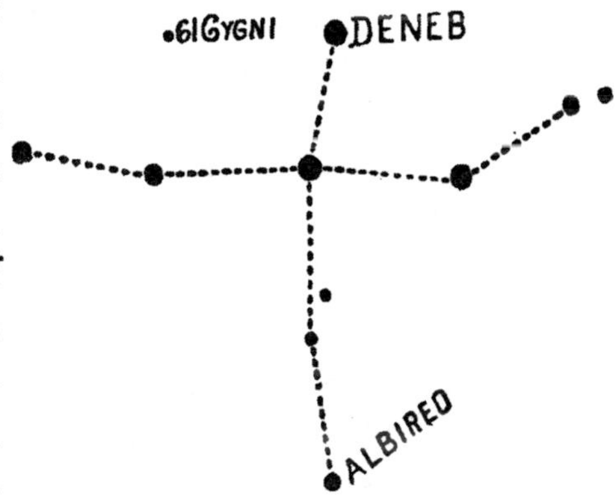

CYGNUS.

Next let us turn to Alpha Centauri—named *Alpha*,

the first letter of the Greek alphabet, because it is the brightest star in the constellation of the Centaur. The second brightest star in a constellation is generally called Beta, the third Gamma, the fourth Delta, and so on; just as if we were to name them A, B, C, D, in order of brightness.

The constellation Centaur lies in the southern heavens, close to the beautiful constellation called the Southern Cross, and is invisible in the northern hemisphere. Of all the stars shining in the heavens round our earth, two only—Sirius and Canopus—show greater brilliancy than Alpha Centauri.

As in the case of 61 Cygni, astronomers were led to attempt the measurement of Alpha Centauri's distance, by noticing how much more distinct were his movements than the movements of other stars, though less rapid both to the eye and in reality than those of 61 Cygni. Alpha Centauri's rate of motion is some thirteen miles each second.

And this, so far as we yet know, is our sun's nearest neighbor in the heavens outside his own family circle.

Strange to say, Alpha Centauri, like 61 Cygni, consists, not of a single star, but of a pair of stars. It is a two-sun system—whether or no surrounded by planets can not be told. We can only reason from what we see to what we do not see. And as God did not form our earth "in vain," and did not form our sun "in vain," so we firmly believe that he did not form "in vain" any one of his myriads of suns scattered through space. What is, has been, or will be the particular use of each one, it would be rash to attempt to say. But that many among them have, like our own sun, systems of worlds revolving round them, we may safely consider very probable.

The two suns of the Alpha Centauri double-star

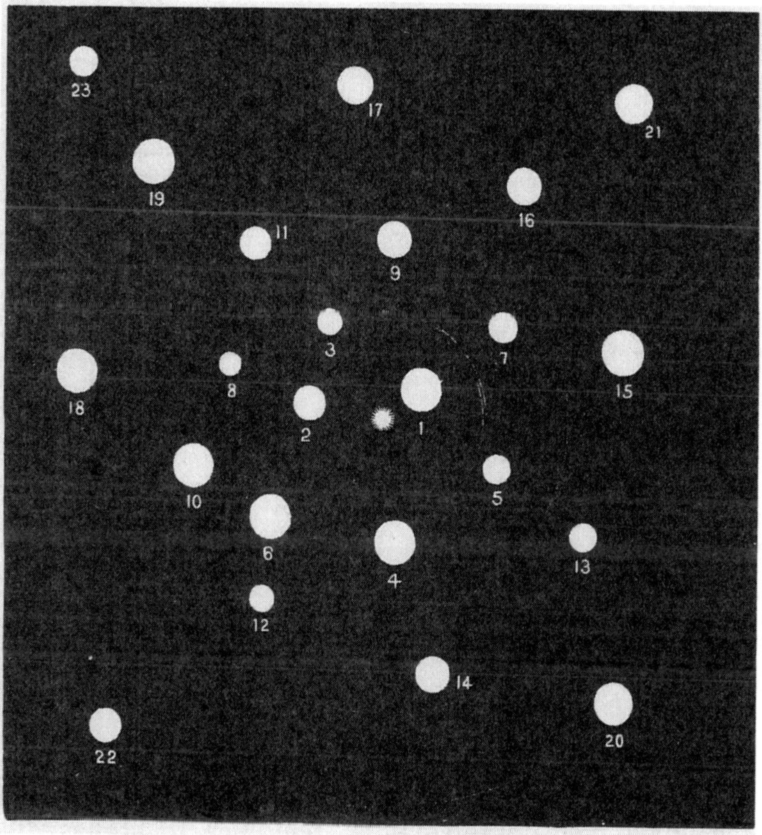

STARS WHOSE DISTANCES ARE BEST KNOWN. (See Table.)

are separated by a distance about twenty-two times as great as the distance of the earth from the sun, yet to the naked eye they show as a single star. Here again one is much smaller than the other; and the smaller revolves round the larger in about eighty-five years.

It is believed that the two together would form a

sun about twice as large and heavy as our sun. This belief is strengthened by the great brilliancy of Alpha Centauri. Our own sun, placed at that distance from us, would shine only one-third as brightly as he does

STARS WHOSE DISTANCES ARE BEST KNOWN.

	NAME OF STAR.	MAGNITUDE.	DISTANCE IN MILES.
1	Alpha Centauri, . . .	1.0	25 trillions.
2	61 Cygni, ...	5.1	43 "
3	2,398, Draco, ...	8.2	55 "
4	Sirius, ...	1.0	58 "
5	9,352, Lacaille, . . .	7.5	66 "
6	Procyon, ...	1.3	71 "
7	Lalande, 21,258, . . .	8.5	74 "
8	Œltzen, 11,677, . . .	9.0	74 "
9	Sigma Draconis, . .	4.7	78 "
10	Aldebaran, ...	1.5	81 "
11	Epsilon Indi, ...	5.2	87 "
12	Œltzen, 17,415, . . .	9.0	94 "
13	1,516, Draco, ...	7.0	101 "
14	Omicron Eridani, . .	4.4	101 "
15	Altair, ...	1.6	101 "
16	Bradley, 3,077, . . .	5.5	101 "
17	Eta Cassiopeiæ, . .	3.6	118 "
18	Vega, ...	1.0	128 "
19	Capella, ...	1.2	174 "
20	Arcturus, ...	1.0	204 "
21	Pole Star, ...	2.1	215 "
22	Mu Cassiopeiæ, . .	5.2	320 "
23	1,830, Groombridge, .	6.5	426 "

This table presents the most trustworthy data which we have yet obtained with reference to stellar distances. As a great number of attempts have been made on stars which, by their brightness or by the magnitude of their proper motion, would appear to be the nearest to us, we may believe that the star now considered as the nearest is really so, and that there is no other less distant. Thus our sun, a star in the immensity, is isolated in infinitude, and *the nearest* sun reigns at twenty-five trillions of miles from our terrestrial abode. Notwithstanding its unimaginable velocity of 186,400 miles a second, light moves, flies, during four years and one hundred and twenty-eight days to come from this sun to us. Sound would take more than three millions of years to cross the same abyss. At the constant velocity of thirty-seven miles an hour, *an express train starting from the sun Alpha Centauri would not arrive here till after an uninterrupted course of nearly seventy-five millions of years.*

Here, then, are the nearest suns to us. These stars, twenty-three in number, are almost the only ones which have shown a perceptible parallax; still the result is very doubtful for the last four, of which the parallax is less than a tenth of a second. Attempts have been made to ascertain the parallax of all the stars of the first magnitude, and the result has been negative for those which are not entered in this list. Canopus, Rigel, Betelgeuse, Achernar, Alpha of the Cross, Antares, Spica, and Fomalhaut, do not show a perceptible parallax. The fine star Alpha Cygni, which shines near 61 Cygni, does not present to the most accurate researches any trace of fluctuation: it is, then, incomparably more distant than its modest neighbor—at least five times, and perhaps twenty times, fifty times, one hundred times beyond that. What must be the colossal size and amazing light of these suns, of which the distance is greater than three hundred to four hundred trillions of miles, and which nevertheless still shine with so splendid a brightness!

We did not know the distance of any stars till the year 1840. This shows how recent this discovery is;

indeed, we have hardly now begun to form an approximate idea of the real distances which separate the stars from each other. The parallax of 61 Cygni, the first which was known, was determined by Bessel, and resulted from observations made at Königsberg from 1837 to 1840. Since then, the first figure obtained has been corrected by a series of more recent observations.

Such distances are amazing and almost terrifying to the imagination. The mind is bewildered and almost overwhelmed when attempting to form a conception of such portions of immensity, and feels its own littleness, the limited nature of its powers, and its utter incapacity for grasping the amplitudes of creation. But though it were possible for us to wing our flight to such a distant orb as 61 Cygni, we should still find ourselves standing only on the verge of the starry firmament where ten thousands of other orbs a thousand times more distant would meet our view. We have reason to believe that a space equal to that which we are now considering intervenes between most of the stars which diversify our nocturnal sky. The stars appear of different magnitudes; but we have the strongest reason to conclude that in the majority of instances this is owing, not to the difference of their real magnitudes, but to the different distances at which they are placed from our globe.

If, then, the distance of a star of the first or second magnitude, or those which are nearest to us, be so immensely great, what must be the distance of stars of the sixteenth or twentieth magnitude, which can be distinguished only by the most powerful telescopes?

Some of these must be many millions of times more distant than the nearest star whose distance now appears to be determined. And what shall we think of the distance of those which lie beyond the reach of any telescope that has yet been constructed, stretching beyond the utmost limits of mortal vision, within the unexplored regions of immensity? Here even the most vigorous imagination drops its wing, and owns itself utterly unable to penetrate this mysterious and boundless unknown.

Turning now from the sun, whose distance was first measured, and from the nearest star with which we are acquainted, let us think about the most radiant star in the heavens—Sirius, "the blazing Dog-star of the ancients;" named by one astronomer "the king of suns."

First, as to the color of Sirius. He belongs to the order of "White Suns," and among all the white suns known to us, Sirius ranks as chief. There may be many at greater distances far surpassing him in size, and weight, and brilliancy; but we can only speak so far as we know. Strange to say, Sirius was not always a "white sun;" for ancient writers describe him as being, in their days, a *red* star. This change is very singular, and difficult to understand. But Sirius is not the only example of the kind. Many alterations in color have been noticed as taking place, often in a much shorter space of time. For instance, one star, which was a white sun in the days of Herschel, is now a golden sun.

Secondly, as to the distance of Sirius. Like a few other stars, Sirius lies not quite so far away as to be beyond reach of measurement. No base-line upon earth would cause the slightest seeming change of position in him; but as our earth journeys round the sun, the line from one side of her orbit to the other is found wide enough. A base-line of one hundred and eighty-six millions of miles does cause just a tiny seeming change.

It is very little even with Alpha Centauri, and with Sirius it is much less. The "displacement" of Sirius is so slight that to measure his distance with exactness is impossible. Sirius lies more than twice as far away from earth as Alpha Centauri. To reach Sirius, the straight line between earth and sun must be repeated six hundred and twenty-six thousand times. Light, which reaches us from the sun in eight minutes and a half, and from Alpha Centauri in four years and a third, can not reach us from Sirius in less than nine years.

Thirdly, as to the size of Sirius. Here, of course, we are in difficulties. Radiantly as Sirius shines on a clear night, and dazzling as he looks through a powerful telescope, he shows no real disk or round surface, but only appears as one point of brilliant light. Some believe him to be much the same size as our sun, while others believe him to be very much larger. But we have no certain ground to rest upon. The only safe method of calculating the probable size—or, rather, weight—of a star is through the discovery of a companion, and a knowledge of its distance from the chief star, and its time of revolution round the chief star.

Fourthly, as to the motions of Sirius—not his nightly apparent movement, caused by the earth's turning on her axis, but his real journey through space. For a long while it was only possible to observe the movements of a star when the star was traveling *sideways* to us across the sky. A star coming straight towards us, or going straight away from us, would seem to be at rest. Lately, however, by means of that remarkable instrument, the spectroscope, it has been found possible to measure the motions of stars coming towards or going away from us. This is possible as yet only in a few scattered cases, but among those few is the brilliant Sirius.

The sideways motion of Sirius had been known before. It now appears that he does not move exactly sideways to us, but is rushing away in a slanting direction at the rate of thirty miles each second, or about one thousand millions of miles a year.

How many millions upon millions of miles Sirius must now be farther from us than in the days of the ancients! Yet he shines still, the most brilliant star in the sky. So small a matter are all those millions of miles, compared with the whole of his vast distance from us, that we do not perceive any lessening of light.

It is an interesting fact that while Sirius is moving away from us, we are also moving away from him. Our "first station ahead," in the constellation Hercules, lies exactly in the opposite direction from Sirius. But the sun does not move so fast as Sirius. He is believed to accomplish only about one hundred and fifty millions of miles each year, drawing all his planets with him.

Fifthly, has Sirius a family, or system, like that of

our sun? Why not? Common sense and reason alike answer with a "Probably, yes"—not only about Sirius, but about other stars also. No doubt the systems—the number, size, weight, speed, distance, and kind of planets—differ very greatly. No doubt there are boundless varieties of beauty and grandeur.

But that our sun should be the head of so wonderful and complex a system, that his rays should be the source of life and heat to so many dependents, and that all the myriads of suns besides should be mere solitary lamps shining into empty space, warming, lighting, controlling *nothing*, is an idea scarcely to be looked in the face.

Of course, there may be other and different uses for some of these suns, beyond our understanding. We must not be positive in the matter. It seems, however, pretty certain that Sirius at least is not a solitary, unattended sun.

Astronomers, carefully watching his movements, were somewhat perplexed. As with Uranus, the attraction of a heavy body beyond was suspected from the nature of the planet's motions, so it happened again with the far more distant Sirius. Astronomers could only explain his movements by supposing the attraction of some large and near satellite. No one knew anything about such a satellite, but it was felt that one must exist.

Nearly twenty years had elapsed after Bessel had predicted the disturber of Sirius before the telescopic discovery which confirmed it was made. The circumstances under which that discovery was made are not, indeed, so dramatic as those which attended the discovery of Neptune, but yet they have an interest of their own. In February, 1862, Alvan Clark & Sons, the celebrated telescope-makers, of Cambridge, Massachusetts, were completing a superb eighteen-inch object-glass for the Chicago Observatory. Turning the instrument on Sirius, for the purpose of trying it, the practiced eye of Alvan Graham Clark, the son, soon detected something unusual, and he exclaimed, "Why, father, the star has a companion!" The father looked, and there was a faint companion due east from the bright star, and distant about ten seconds of a degree. This was exactly the predicted direction of the companion of Sirius, and yet the observers knew nothing of the prediction. As the news of this discovery spread, many great telescopes were pointed on Sirius; and it was found that when observers knew exactly where to look for the object, many instruments would show it. The new companion star to Sirius lay in the true direction, and it was now watched with the keenest interest, to see whether it also was moving in the way it should move,—if it were really the body whose existence had been foretold. Four years of observation showed that this was the case, so that hardly any doubt could remain that the telescopic discovery had been made of the star which had caused the inequality in the motion of Sirius.

It is certainly a very remarkable fact that, out of the thousands of stars with which the heavens are adorned, no single star has yet been found which certainly shows an appreciable disk in the telescope. We are aware that some skillful observers have thought that certain small stars do show disks; but we may lay this aside, and appeal only to the ordinary fact that our best telescopes turned on the brightest stars show merely glittering points of light, so hopelessly small as to elude our most delicate micrometers. The ideal astronomical telescope is, indeed, one which will show Sirius, or any other bright star, as nearly identical as possible with Euclid's definition of a point, being that which has no parts and no magnitude.

Lastly, what is Sirius made of? The late discovery of spectrum analysis, or of the instrument called the spectroscope, helps us here. A little further explanation on this subject will be given later. It may be observed in passing that before the said discovery, astronomers can scarcely be asserted to have known with any certainty that stars were suns. They could, indeed, be sure that at the vast distances of the fixed stars no bodies, shining by merely reflected light, could possibly be visible to us. But there knowledge stopped.

Now we can say more. Now the spectroscope, by breaking up and dividing for us the slender ray of light traveling from each star, has shown to us something of the nature of the stars. Now we know that the stars, like our sun, are burning bodies, surrounded by atmospheres full of burning gas.

It has even been found out that in Sirius there are large quantities of sodium and magnesium, besides other metals, and an abundant supply of hydrogen gas.

CHAPTER XXIV.

DIFFERENT KINDS OF SUNS.

VARIOUS in kind, various in size, various in color, various in position, various in motion, are the myriad suns scattered through space. So far are they from being formed on the same plan, turned out on the same model, that it may with reason be doubted whether any two stars could be found exactly alike. Why should we expect to find them so? No two oak-leaves, no two elm-leaves, precisely alike, are to be found upon earth.

So some stars are large, some are small. Some are rapid in movement, some are slow. Some are yellow, some white, some red, green, blue, purple, or gray. Some are single stars; others are arranged in pairs, trios, quartets, or groups. Some appear only for a time, and then disappear altogether. Others are changeful, with a light that regularly waxes and wanes in brightness.

We have now to give a little time and thought to variable stars and temporary stars; afterwards to double stars and colored stars. There are many stars which pass through gradual and steady changes—first brightening, then lessening in light, then brightening again. One such star is to be seen in the constellation of the Whale. It is named "Mira," or "The Marvelous," and the time in which its changes take place extends to eleven months. For about one fortnight it is a star of the second magnitude. Through three months it grows slowly more and more dim, till it becomes invisible, not only to the naked eye, but through ordinary telescopes. About five months it remains thus. Then again, during three months, it grows brighter and brighter, till it is once more a second-magnitude star, and after a fortnight's pause begins anew to fade.

At its maximum, this star is yellow; when it is faint, it is reddish. Spectrum analysis shows in it a striped spectrum of the third type, and when its light diminishes it preserves all the principal bright rays reduced to very fine threads. The most plausible explanation of this variability is to suppose that it periodically emits vapors similar to the eruptions observed in the solar photosphere. Instead of its periodicity being like that of the sun—eleven years and hardly perceptible—this variation of the sun of the Whale is three hundred and thirty-one days and very considerable. It is subject to oscillations and irregularities similar to those which we remark in our sun. Of all the variable stars, this is the easiest to observe, and it has been known for nearly three hundred years.

The most celebrated of all the variable stars is that known as Algol, in the constellation of Perseus. This star is very conveniently placed for observation, being visible every night in the Northern Hemisphere, and its wondrous and regular changes can be observed without any telescopic aid. Every one who desires to become acquainted with the great truths of astronomy should be able to recognize this star, and should have also followed it during one of its periods of change. Algol is usually a star of the second magnitude; but in a period between two or three days—or, more accurately, in an interval of two days, twenty hours, forty-eight minutes, and fifty-five seconds—its brilliancy goes through a most remarkable cycle of variations. The series commences with a gradual decline of the star's brightness, which, in the course of three or four hours, falls from the second magnitude down to the fourth. At this lowest stage of brightness, Algol remains for about twenty minutes, and then begins to increase, until in three or four hours it regains the second magnitude, at which it continues for about two days, thirteen hours, when the same series commences anew. It seems that the period required by Algol to go through its changes is itself subject to a slow, but certain variation.

Another variable star, Betelgeuse in Orion, undergoes its variations in about two hundred days; while yet another, Delta Cephei, takes only six days. Our own sun is believed to be in some slight degree a variable star, passing through his changes in eleven years. When the sun-spots are most numerous, he would probably appear, if seen from a great distance, more dim than when there are few or none of them.

Sometimes a star is not merely variable. Stars have appeared for a brief space, and then utterly vanished. They are then named Temporary Stars. Whether such a star really does go out of existence, or whether it merely becomes too dim to be seen; whether the furnace-fires are extinguished, and the glowing sun changes into a dark body, or whether it merely turns a dark side towards us for a very long period,—these are questions we can not answer.

An extraodinary specimen of a temporary star was seen in 1572. It was not a comet, for it had no coma or tail, and it never moved from its place. The brightness of the star was so great as to surpass Sirius and Jupiter, and to equal Venus at her greatest brilliancy. Nay, it must have surpassed even Venus, for it was plainly visible at midday in a clear sky. Gradually the light faded and grew more dim, till at length it entirely disappeared. As it lessened in brilliancy, it also changed in color, passing from white to yellow, and from yellow to red. This curiously agrees with the three tints of the first, second, and fourth orders of suns, as lately classified.

This star was observed by Tycho Brahe, who gives the following curious account:

"One evening, when I was contemplating, as usual, the celestial vault, whose aspect was so familiar to me, I saw, with inexpressible astonishment, near the zenith, in Cassiopeia, a radiant star of extraordinary magnitude. Struck with surprise, I could hardly believe my eyes. To convince myself that it was not an illusion, and to obtain the testimony of other persons, I called out the workmen employed in my laboratory, and asked them, as well as all passers-by, if they could see, as I did, the star, which had appeared all at once. I learned later on that in Germany carriers and other people had anticipated the astronomers in regard to a great apparition in the sky, which gave occasion to renew the usual railleries against men of science (as with comets whose coming had not been predicted).

"The new star was destitute of a tail; no nebulosity surrounded it; it resembled in every way other stars of the first magnitude. Its brightness exceeded that of Sirius, of Lyra, and of Jupiter. It could only be compared with that of Venus when it is at its nearest possible to the earth. Persons gifted with good sight could distinguish this star in daylight, even at noonday, when the sky was clear. At night, with a cloudy sky, when other stars were veiled, the new star often remained visible through tolerably thick clouds. The distance of this star from the other stars of Cassiopeia, which I measured the following year with the greatest care, has convinced me of its complete immobility. From the month of December, 1572, its brightness began to diminish; it was then equal to Jupiter. In January, 1573, it became less brilliant than Jupiter; in February and March, equal to stars of the first order; in April and May, of the brightness of stars of the second order. The passage from the fifth to the sixth magnitude took place between December, 1573, and February, 1574. The following month the new star disappeared without leaving a trace visible to the naked eye, having shone for seventeen months."

These circumstantial details permit us to imagine the influence which such a phenomenon must have exercised on the minds of men. Few historical events have caused so much excitement as this mysterious envoy of the sky. It first appeared on November 11, 1572. General uneasiness, popular superstition, the fear of comets, the dread of the end of the world, long since announced by the astrologers, were an excellent setting for such an apparition. It was soon announced that the new star was the same which had led the wise men to Bethlehem, and that its arrival foretold the return of the Messiah and the last judgment. For the hundredth time, perhaps, this sort of prognostication was recognized as absurd. It did not, however, prevent the astrologers from being believed, twelve years later, when they announced anew the end of the world for the year 1588; these predictions exercised the same influence on the public mind.

After the star of 1572 the most celebrated is that which appeared in October, 1604, in Serpentarius, and which was observed by two illustrious astronomers, Kepler and Galileo. As happened with the preceding, its light imperceptibly faded. It remained visible for fifteen months, and disappeared without leaving any traces. In 1670 another temporary star, blazing out in the head of the Fox (Vulpecula), showed the singular phenomenon of being extinguished and reviving several times before it completely vanished. We know of *twenty-four* stars which, during the last two thousand years, have presented a sudden increase of light; have been visible to the naked eye, often brilliant; and have then again become invisible. The last apparitions of this kind happened before our eyes in 1866, 1876, 1885, and 1892, and spectrum analysis enabled us to ascertain, as we have seen, that they were due to veritable combustion—a fire caused by a tremendous expansion of incandescent hydrogen, and to phenomena analogous to those which take place in the solar photosphere. A rather curious fact about these stars is, that they do not blaze out indifferently in any point of the sky, but in rather restricted regions, chiefly in the neighborhood of the Milky Way.

Many other instances have been known, beside those referred to, of variable and temporary stars.

There are two distinct kinds of double-stars. First, we have those which merely seem to be double, because one lies almost directly behind the other, though widely distant from it. Just as a church-tower, two miles off, may appear to stand close side by side with another church-tower two and a half miles off, though they are in fact separated. Secondly, we have the real systems of two suns belonging to one another; the smaller moving round the larger, or more correctly both traveling round one central point called the center of gravity, the smaller having the quicker rate of motion.

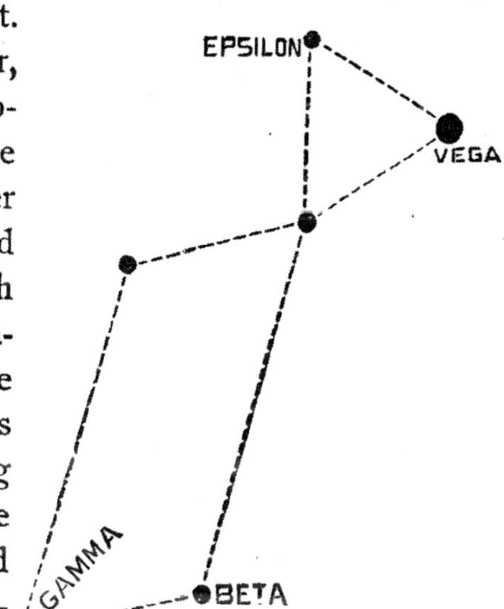

LYRA.

Alpha Centauri and 61 Cygni have been already described, as examples of true double-stars. In the constellation Lyra a marked instance is to be seen. The brightest star in the Lyre is Vega, and near Vega shines a tiny star, which to people with particularly clear sight has sometimes rather a longish look. If you examine this star through an opera-glass, you find it to consist of two separate stars. But if you get a more powerful telescope, and look again, you will find that *each star* of the couple actually consists of *two* stars. The four are not at equal distances. Two points of light seemingly close together are parted by a wide gap from two other points equally close together. These four stars are believed to have a double motion. Each of the separate pairs revolves by itself, the two suns traveling round one center; and in addition to this the two *couples* of suns probably perform a long journey round another center common to them all.

Many thousands of double stars have been discovered; and a large number of these are now known to be, not merely two distinct suns lying in the same line of sight, but two brother-suns, each probably the center of his own system of planets.

We have not only to consider the number of suns, though of simple numbers more yet remains to be said. Attention must also be given to the varying colors of different stars; for all suns in the universe are not made after the model of our sun. All suns are not yellow.

So far as single stars are concerned, colors seem rather limited. White stars, golden or orange stars, ruby-red stars, placed alone, are often seen; but blue stars, green stars, gray stars, silver stars, purple stars, are seldom if ever visible to the naked eye, or known to exist as single stars.

Take a powerful telescope and examine star-couples, and a very different result you will find. Not white, yellow, and red alone, but blue, purple, gray, green, fawn, buff, silvery white, and coppery hues, will delight you in turn. As a rule, when the two stars of a couple are alike in color, they are either white, or yellow, or red. Also in the case of double-stars of different colors, the larger of the two is almost invariably white or some shade of yellow or red.

There are, however, exceptions to all such rules. Blue stars are almost never seen alone, and as one of a pair the blue star is generally, if not invariably, the smaller. But instances are known of double-stars, both of which are blue; and one group in the southern heavens is entirely made up of a multitude of bluish suns.

It is when we come to consider double-stars of two colors that the most striking effects are found. Now and then the two suns are nearly the same in size; but more commonly one is a good deal larger than the other. This is known by the brighter light of the largest, and the more rapid movements of the smallest. The lesser star is often only small by comparison, and may be in reality a very goodly and brilliant sun.

Among nearly six hundred "doubles" examined by one astronomer, there were three hundred and seventy-five in which the two stars were of one color, generally white, yellow, orange, or red. The rest were different in tint, the difference between the two suns in about one hundred and twenty cases being very marked.

For instance, a red "primary," as the larger star is called, will be seen with a small green satellite; or a white primary will have a little brother-sun of purple, or of dark ruby, or of light red. Sometimes the larger sun is orange, the companion being purple or dark blue. Again, the chief star will be red with a blue satellite, or yellow with a green satellite, or orange with an emerald satellite, or golden with a reddish-green satellite. We hear of golden and lilac couples,

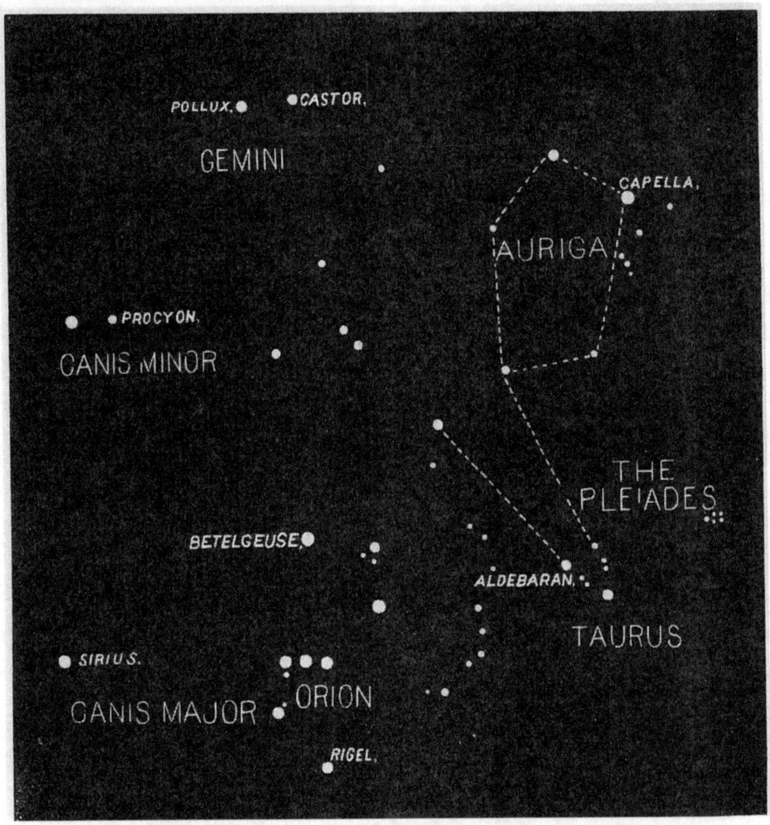

CONSTELLATIONS IN THE NORTHERN HEMISPHERE.

ot cream and violet pairs, of white and green companions. But, indeed, the variety is almost endless.

There may be worlds circling around these suns—worlds, perhaps, with living creatures on them. We know little about how such systems of suns and worlds may be arranged. Probably each sun would have his own set of planets, and both suns with their planets would travel round one central point. Perhaps, where the second sun is much the smallest, it might occasionally be like a big blazing satellite among the planets—a kind of burning Jupiter-sun to the chief sun.

Among colored stars, single and double, a few may be mentioned by name as examples. Sirius, as already observed, is a brilliant white sun; and brilliant white also are Vega, Altair, Regulus, Spica, and many others. Capella, Procyon, the Pole-star, and our own sun, are examples of yellow stars. Aldebaran, Betelgeuse, and Pollux, are ruby-red. Antares is a red star, with a greenish "scintillation" or change of hue in its twinkling. A tiny green sun belonging to this great and brilliant red sun has been discovered. Some have called Antares "the Sirius of Red Suns."

It is during the fine nights of winter that the constellation Taurus, or The Bull, with the star Aldebaran marking its eye, shines in the evening above our heads. No other season is so magnificently constellated as the months of winter. While nature deprives us of certain enjoyments in one way, it offers us in exchange others no less precious. The marvels of the heavens present themselves from Taurus and Orion in the east to Virgo and Boötes on the west. Of eighteen stars of the first magnitude, which are counted in the whole extent of the firmament, a dozen are visible from nine o'clock to midnight, not to mention some fine stars of the second magnitude, remarkable nebulæ, and celestial objects well worthy of the attention of mortals. It is thus that nature establishes an harmonious compensation, and while it darkens our short and frosty days of winter, it gives us long nights enriched with the most opulent creations of the sky. The constellation of Orion is not only the richest in brilliant stars, but it conceals for the initiated treasures which no other is known to afford. We might almost call it the California of the sky.

To the southeast of Orion, on the line of the Three Kings, shines the most magnificent of all the stars, *Sirius*, or *Alpha* of the constellation of the Great Dog. This constellation rises in the evening at the end of November, passes the meridian at midnight at the end of January, and sets at the end of March. It played the greatest part in Egyptian astronomy, for it regulated the ancient calendar. It was the famous Dog Star; it predicted the inundation of the Nile, the summer solstice, great heats and fevers; but the precession of the equinoxes has in three thousand years moved back the time of its appearance by a month and a half; and now this fine star announces nothing, either to the Egyptians who are dead or to their successors. But we shall see farther on what it teaches us of the grandeur of the sidereal universe.

The *Little Dog*, or Procyon, is found above the Great Dog and below the Twins (Castor and Pollux), to the east of Orion. With the exception of Alpha Procyon, no brilliant star distinguishes it.

The two double-stars, 61 Cygni and Alpha Centauri, are formed each of two orange suns.

In the Southern Cross there is a wonderful group of stars, consisting of about one hundred and ten suns, nearly all invisible to the naked eye. Among the principal stars of this group, which Sir John Herschel described as being, when viewed through a powerful telescope, like "a casket of variously-colored precious stones," are two red stars, two bright green, three pale green, and one of a greenish blue.

The splendor of these natural illuminations can hardly be conceived by our terrestrial imagination. The tints which we admire in these stars from here can give but a distant idea of the real value of their colors. Already, in passing from our foggy latitudes to the limpid regions of the tropics, the colors of the stars are accentuated, and the sky becomes a veritable casket of brilliant gems. What would it be if we could transport ourselves beyond the limits of our atmosphere? Seen from the moon, these colors would be splendid. Antares, Alpha Herculis, Pollux, Aldebaran, Betelgeuse, Mars, shine like rubies; the Polar Star, Capella, Castor, Arcturus, Procyon, are veritable celestial topazes; while Sirius, Vega, and Altair are diamonds, eclipsing all by their dazzling whiteness. How would it be if we could approach the stars so as to perceive their luminous disks, instead of merely seeing brilliant points destitute of all diameter?

Blue days, violet days, dazzling red days, livid green days! Could the imagination of poets, could the caprice of painters, picture on the palette of fancy a world of light more astounding than this? Could the mad hand of the chimera, throwing on the receptive canvas the strange lights of its fancy, erect by chance a more astonishing edifice?

CHAPTER XXV.

GROUPS AND CLUSTERS OF SUNS.

We have been thinking a good deal about single stars and double stars, as seen from earth. Now we have to turn our attention to groups, clusters, *masses* of stars, in the far regions of space.

Have you ever noticed on a winter night, when the sky was clear and dotted with twinkling stars, a band of faintly-glimmering light stretching across the heavens from one horizon to the other? The band is irregular in shape, sometimes broader, sometimes narrower; here more bright, there more dim. If you were in the Southern Hemisphere you would see the same soft belt of light passing all across the southern heavens. This band or belt is called the Milky Way.

But what is the Milky Way? It is made up of stars. So much we know. As the astronomer turns his telescope to the zone of faintly-gleaming light, he finds stars appearing behind stars in countless multitudes; and the stronger his telescope, the more the white light changes into distant stars.

Our sun we believe to be one of the stars of the Milky Way; merely one star among millions of stars; merely one golden grain among the millions of sparkling gold-dust grains scattered lavishly through creation. Scattered, not recklessly, not by chance, but placed, arranged, and guided each by its Maker's upholding hand.

The Milky Way, or the Galaxy, as it has been called, has great interest for astronomers. Many have been the attempts made to discover its actual size, its real shape, how many stars it contains, how far it extends; but to all such questions the only safe answers to be returned are fenced around with "perhaps" and "may be." There are many very remarkable clusters of

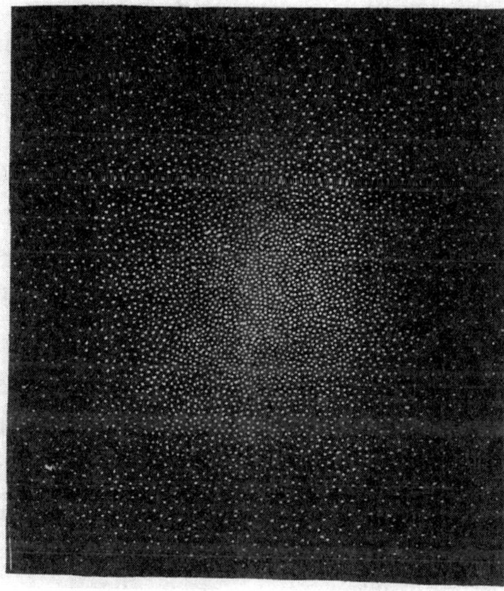

A CLUSTER OF STARS IN CENTAURUS.

stars to be seen in the heavens—some few visible as faint spots of light to the naked eye, though the greater number are only to be seen through a telescope. Either with the naked eye, or in telescopes of varying power, they show first as mere glimmers of light, which, viewed with a more powerful telescope, separate into clusters of distant stars.

The most common shape of these clusters is globular—to the eye appearing simply round. Stars gather densely near the center, and gradually open out to a thin scattering about the edge. Thousands of suns are often thus collected into one cluster.

The clusters are to be seen in all parts of the sky; but the greater number seem to be gathered into the space covered by the Milky Way and by the famous south Magellanic Clouds.

Some of them are beautifully colored; as, for instance, a cluster in Toucan, not visible from England, the center of which is rose-colored, bordered with white. No doubt it contains a large number of bright red suns, surrounded by a scattering of white suns.

It used to be supposed that many of these clusters were other vast gatherings or galaxies of stars, like the Milky Way, lying at enormous distances from us. This now seems unlikely. The present idea rather is, that each cluster, in place of being another "Milky Way" of millions of widely-scattered suns, is one great *star-system*, consisting indeed of thousands of suns, but all moving round one center. Probably most of these clusters are themselves a part of the collection of stars to which our sun belongs.

If this be so, and if worlds are traveling among the suns—as may well be, since they are doubtless quite far enough apart for each sun to have his own little or great system of planets—what sights must be seen by the inhabitants of such planets!

We do not indeed know the distance between the separate suns of a cluster, which may be far greater than appears to us. But if astronomers calculate rightly, they are near enough together to shed bright light on all sides of a planet revolving in their midst. The said planet might perhaps not have within view a single sun equal in apparent size to our sun as seen from earth; yet thousands of lesser suns, shining brightly in the firmament night and day, would cause a radiance which we never enjoy.

No, not *night*. In such a world there could be no night. Worlds in the midst of a star-cluster must be

THE GREAT NEBULA IN ANDROMEDA, COMPARED WITH SIZE OF THE SOLAR SYSTEM.

regions of perpetual day. No night, no starry heaven, no sunrise lights or sunset glories, no shadow mingling with sunshine, but one continual, ceaseless blaze of brightness. We can hardly picture, even in imagination, such a condition of things.

Besides star-clusters there are also nebulæ. The word *nebula* comes from the Latin word for "cloud," and the nebulæ are so named from their cloudlike appearance.

It is not easy to draw a line of clear division between nebulæ and distant star-clusters; for both have at first sight the same dim, white, cloudy look. In past days the star-clusters were included by astronomers under the general class of nebulæ. And as with the star-clusters, so with many of the nebulæ, the more powerful telescopes of modern days have shown them to be great clusters, or systems, or galaxies of stars, at vast distances from us.

It was supposed with the nebulæ, as with some of the star-clusters, that they were other "Milky Ways" of countless stars, far beyond the outside boundaries of our Milky Way. This may still be true of some or many nebulæ, but certainly not of all.

For there are different kinds of nebulæ. Some may, as just said, be vast gatherings of stars lying at distances beyond calculation, almost beyond imagination. Others appear rather to be clusters of stars, like those already described, probably situated in our own galaxy of stars. There is also a third kind of nebula. Many nebulæ, once supposed to be clusters of stars, having been lately examined by means of the spectroscope, are found to be enormous masses of glowing gas, and not solid bodies at all.

The number of nebulæ known amounts to many thousands. They are commonly divided into classes, according to their seeming shape. There are nebulæ of regular form, and nebulæ of irregular form. There are circular nebulæ, oval nebulæ, annular nebulæ, conical nebulæ, cometary nebulæ, spiral nebulæ, and nebulæ of every imaginable description. These shapes would no doubt entirely change, if we could see them nearer; and indeed, even in more powerful telescopes, they are often found to look quite different. While on the subject of clusters and nebulæ, mention should be made of the famous Magellanic Clouds in the southern heavens. Sometimes they are called the Cape Clouds. They differ from other nebulæ in many points, and more particularly in their apparent size. The Great Cloud is about two hundred times the size of the full moon, while the Small Cloud is about one-quarter as large. In appearance they are not unlike two patches of the Milky Way, separated and moved to a distance from the main stream.

These clouds are surrounded by a very barren portion of the heavens, containing few stars; but in themselves they are peculiarly rich. Seen through a powerful telescope, they are found to abound with stars. The Greater Cloud alone contains over six hundred from the seventh to the tenth magnitudes, countless tiny star-points of lesser magnitudes, star-clusters of all descriptions, and nearly three hundred nebulæ,—all crowded into this seemingly limited space.

Among many famous nebulæ, the one in Orion and the one in Andromeda may be particularly mentioned. The size of the latter, as compared with the size of our Solar System, is shown in the cut of this nebula. On the hypothesis of a complete resolvability into stars, the mind is lost in numbering the myriads of suns, the agglomerated individual lights of which produce these nebulous fringes of such different intensities. What must be the extent of this universe, of which each sun is no more than a grain of luminous dust!

CHAPTER XXVI.

THE PROBLEM OF SUN AND STAR DISTANCES.

One question had long occupied the minds of scientific men without the finding of any satisfactory answer, and this was the distance of the sun from our earth.

At length the sun had fully gained his rightful position in the minds of men as center and controller of the Solar System, and earth was fully dethroned from her old false position. In point of fact, the sun had gained *more* than his rightful position; since he was now looked upon very much as the earth had been formerly looked upon; since he was regarded practically as the motionless Universe-Center. The earth might and did move—people had grown used to that idea—but the sun, at all events, was fixed. The sun, beyond turning upon his axis, was motionless.

Still, as to the true distance of the sun from ourselves, ignorance reigned. Tycho Brahe had made a great advance on earlier notions by placing the light-giver 4,000,000 miles away in imagination. Kepler, a quarter of a century later, increased this to 14,000,000. Galileo afterwards reverted to the 4,000,000. All this, however, was sheer guesswork.

Not till well on in the seventeenth century did Cassini make a definite attempt at actual measurement of the sun's distance, and this attempt gave as a result some 82,000,000 miles—not far removed from the truth. But at the same period other observers gave results of 41,000,000 and 136,000,000; so for years the question still remained swathed in mist. Better modes and better instruments were required before it could receive settlement.

The measurement of the distance of objects a good way off is by means of their "parallax," or the apparent change in their position when viewed from two different points at some distance from each other. To this method of calculating distance we have already referred.

This kind of measurement of distances is not confined to our earth's surface alone. The distances of bright bodies in the heavens may be calculated after just the same method, provided only that the heavenly body is not too far distant to be made to change its seeming position in the sky.

Suppose you wished to find out how far off the moon is. You would have to make your observation of the moon's exact position in the sky—of just that spot precisely where she is to be seen among heavenly scenery at a particular moment; and you would have to get somebody else to make a similar observation, at the same time, from some other part of earth, at a good distance from your post.

There are no hills or woods in the sky for the moon to be seen against; but there are plenty of stars, bright points of light far beyond the moon. If you look at her from one place, and your friend looks at her from another place a good way off, there will be a difference in your two "views." You will see her very close to certain stars; and your friend will see her not quite so close to those stars, but closer to others. This difference of view would give the moon's parallax.

If your observation were very careful, very exact, and if you had the precise distance between the spot where you stood, and the spot where your friend stood, —then, from that base-line, and from the two angles formed by its junction with the lines from the two ends of it straight to the moon, you might reckon how many miles away the moon is. The greater the distance between the two places of observation, the better,—as, for instance, at Greenwich and the Cape of Good Hope.

Practically, this is not nearly so simple a matter as it may sound, because our earth is always on the move, and the moon herself is perpetually journeying onward. All such motions have to be most scrupulously allowed for. So the actual measurement of moon-distance requires a great deal of knowledge and of study. Many other difficulties and complications besides these enter into the question, and have to be overcome.

Now, the sun is very much farther away than the moon, and the stumbling-blocks in the way of finding out his distance become proportionately greater and more numerous. To observe him from two parts of England, or from two parts of Europe, would give him no parallax. That is to say, there would be no change of position on his part apparent to us.

I do not say that there would be no change at all; but only that it would be so minute as to be quite unseen by human eyes, even with the help of most careful and accurate measurement. A very long base-line is needed to make the sun distinctly appear to change his position ever so little. Only the very longest base-line which can be found on earth will do for this,—nothing less than the earth's whole diameter of nearly eight thousand miles. And I think you will see that the success of the calculation would then depend, not alone on most careful observation from two posts at the opposite sides of earth; not alone on mathematical gifts and powers of close reckoning; but also, essentially, on a true knowledge of our earth's diameter; that is, of *the exact length of the base-line* from which the whole calculation would have to be made, and upon which the answer would largely depend.

For a long while the earth's diameter was not well known. As time went on, fresh measurements were again and again made of different portions of earth's surface, fresh calculations following therefrom; and gradually clear conclusions were reached. A very important matter it was that they should be reached; for the semi-diameter of our earth has been adopted as a "standard measure" for the whole universe; and the slightest error in that standard measure would affect all after calculations.

When actual observations of sun-distance came to be made, innumerable difficulties arose. Foremost stood the huge amount of that distance. This made precise observations more difficult; and at the same time it made every mistake in observation so much the worse. A little mistake in observing the moon might mean only a hundred miles or so wrong in the answer; but a mistake equally small in observing the sun would lead to an error of many millions of miles.

Again, to observe the sun's exact position among the stars, as with the moon, was not possible; because when the sun is visible, the stars are not visible. Then, too, the dazzling brightness of the sun balked the needed exactitude.

Halley had a brilliant thought before he died. He could not carry it out himself, but he left it to others as a legacy; that was, by observing the transit of Venus from two different points on the earth's surface. At his suggestion, on the first opportunity—which was not till after his death—the above mode was tried of measuring the sun's distance.

A certain observer, stationed on one part of earth, saw the tiny dark body of Venus take one particular line across the sun's bright face. Another observer, standing on quite another and a far-off part of earth, saw the little dark body take quite another line across the sun's bright face. Not that there were two little dark bodies, but that the one body was seen by different men in different places.

From these separate views of the path of Venus across the face of the sun, in connection with what was already known of the earth's diameter, and therefore of the length of base-line between the two places, the distance of the sun was reckoned to be about 95-000,000 miles.

Since that date many fresh attempts have been made, and errors have been set right. Mercury as well as Venus has been used in this matter, and other newer modes of measurement have also been successfully tried. We know now, with tolerable accuracy, that the sun's greater distance from earth is between

92,000,000 and 93,000,000 miles. A curious illustration of sun-distance has been offered by one writer. Sound and light, heat and sensation, all require *time* for journeying. When a child puts his finger into a candle-flame, he immediately shrieks with pain. Yet, quickly as the cry follows the action, his brain is not really aware of the burn until a certain interval has elapsed. True, the interval is extremely minute; still it is a real interval. News of the burn has to be telegraphed from the finger, through the nerves of the arm, up to the brain; and it occupies time in transmission, though so small a fraction of a second that we can not be conscious of it.

Now, try to imagine a child on earth with an arm long enough to reach the sun. His fingers might be scorched by the raging fires there, while yet his brain on earth would remain quite unaware of the fact for about one hundred and thirty years. All through those years, sensation would be darting along the arm exactly as fast as it darts from the finger-tips of an ordinary child on earth to that child's brain.

If the scorching began before he was one year old, he would have become a very aged man, one hundred and thirty years old, before he could know in his mind what was happening in the region of his hand. Moreover, if, on receiving the intimation, he should decide to withdraw his hand from that unpleasantly-hot neighborhood, another hundred and thirty years would elapse before the fingers could receive and act upon the message, telegraphed from the brain, through the nerves of the arm. So much for the distance of the sun from our earth!

But the stars! How far off are the stars?

The distance of the moon is a mere nothing. The distances of the planets have been found out. The distance of the sun has been measured. But the stars—those wondrous points of light, twinkling on, night after night, century after century, unchanged in position save by the seeming nightly pilgrimage of them all across the sky in company, caused only by our earth's restless, continual whirl,—

What about the distances of the stars? Can we measure their distance by means of their parallax? This mode of measurement was tried successfully on the moon, on the planets, and on the sun. But when it was tried upon the stars, from two stations as far apart as any two stations on earth could be, the attempt was a failure. Not a ghost of parallax could be detected with any one star. Not the faintest sign of displacement was seen in the position of a single star when most critically and carefully examined.

Then arose a brilliant thought! What of earth's yearly journey round the sun? If a base-line of 8,000 miles were not enough, compared with the great distance of the stars, this at least remained. Our earth at midsummer is somewhere about 185,000,000 miles away from where she is at midwinter, comparing her position with that of the sun, and reckoning him to be at rest.

In reality, the sun is not at rest, but is in ceaseless motion, carrying with him, wherever he goes, the whole Solar System with as much ease as a train carries its passengers. Those passengers are truly in motion, yet, with regard merely to the train, they are at rest. So each member of the Solar System—attached to the sun, not by Ptolemaic bars, but by the bond of gravity—is borne along by him through space; yet, with respect to each member of that system, the central sun is always and absolutely at rest.

At one time of the year, our earth is on one side of the sun, over 92,000,000 miles distant. Six months later, the earth is on the other side of the sun, not quite 93,000,000 miles away from him in that opposite direction. Twice 93,000,000 comes to 186,000,000. This line, therefore—the diameter of the earth's whole yearly orbit, may be roughly stated as about 185,000,000 of miles in length.

Here surely was a base-line fit to give parallax to any star—or, rather, to make parallax visible in the case of any star. For it is, after all, a question, not of fact, but of visibility; not of whether the thing *is*, but of whether we are able to *see* it.

As the earth journeys on her annual tour round the sun, following a slightly elliptic pathway, the diameter of which is about 185,000,000 miles, each star in the sky must of necessity undergo a change of position, however minute, performing a tiny apparent annual journey in exact correspondence with the earth's great annual journey.

The question is not, Does the star do this? but, Can we *see* the star do this? Its apparent change of position may be so infinitesimal, through enormous distance, that no telescopes or instruments yet made by man can possibly show it to us.

The star-motion of which I am now speaking is purely a seeming movement, not real. Be very clear on this point in your mind. *Real* star-movements, though suspected earlier, were not definitely surmised —one may even say "discovered"—until the year 1718, by Halley. And though Cassini in 1728 referred to this discovery of Halley's, yet very little was heard about the matter until the days of Herschel. Only *apparent* star-motions were generally understood and accepted.

The first and simplest of such seeming star-motions is one which we can all see—the nightly journey of the whole host of stars, caused by our earth's whirl upon her axis.

The second is also simple, but by no means also easily seen. Astronomers reasoned out the logical necessity for such an apparent motion long before it could be perceived. As far back as the days of Copernicus it was felt that if the Copernican System were true, if the earth in very deed traveled round the sun, then the stars ought to change their positions in the sky when viewed from different parts of earth's annual journey. Observations were taken, divided by six months of time and by one hundred and eighty-five millions ot miles of space. And the stars stirred not!

Stupendous as was this base-line, it proved insufficient. So much *more* stupendous was the distance of the stars that the base-line sank to nothing, and once again parallax could not be detected. Not that the seeming change of position in the star did not take place, but that human eyes were unable to see it, human instruments were unable to register it.

There lies the gist of the matter. If the change

of position can be observed, well and good! The length of the base-line being known, the distance of the star may be mathematically calculated. For the size of the tiny apparent path, followed in a year by the star, is and must be exactly proportioned to the distance of the star from that base-line. If the tiny oval be so much larger, then the star is known to be so much nearer. If the tiny oval be so much smaller, then the star is known to be so much farther away.

But when not the minutest token could be discovered of a star's position in the sky being in the least degree affected by the earth's great annual change of position, astronomers were at a loss. There was absolutely nothing to calculate from. The star was a motionless point to earth. The whole yearly orbit of earth was a motionless point to the star. One slender beam of light united the two. Reckoners had nothing to stand upon.

It is interesting to know that Copernicus had actually, long before, suggested this as a possible explanation of the absence of star-parallax. He thought that astronomers might fail altogether to find it, because the stars might be "at a practically infinite distance" from our earth.

CHAPTER XXVII.

MEASUREMENT OF STAR DISTANCES.

UNTIL the days of Sir William Herschel, little attention was bestowed upon the universe of distant stars. Before his advent the interest of astronomers had been mainly centered in the sun and his revolving worlds. The "fixed stars" were indeed studied, as such, with a certain amount of care, and numberless efforts were made to discover their distances from earth; but the thought of a vast Starry System, in which our little Solar System should sink to a mere point by comparson with its immensity, had not yet dawned.

With larger and yet larger telescopes of his own making, Herschel studied the whole Solar System, and especially the nature of sun-spots, the rings of Saturn, the various motions of attendant moons, together with countless other details. He enlarged the system by the discovery of another planet outside Saturn—the planet Uranus, till then unknown.

These were only first steps. The work which he did in respect of the Solar System was as nothing compared with the work which he did among the stars. His work in our system was supplementary to other men's labors; his work among the stars was the beginning of a new era. Like many before him, he, too, sought eagerly to find star-parallax, and he, too, failed. Not yet had instruments reached a degree of finish which should permit measuring operations of so delicate a nature.

Although he might not detect star-parallax, he sprang a mine upon the older notions of star-fixity. He shook to its very foundations the sidereal astronomy of the day,—the theory, long held, of a motionless universe, motionless stars, and a rotating but otherwise motionless sun. He did away with the mental picture, then widely believed in, of vast interminable fixity and stillness, extending through space, varied only by a few little wandering worlds. As he swept the skies, and endeavored to gauge the fathomless depths, and vainly pursued the search for the parallax of one star after another, he made a great and unlooked-for discovery. This was in connection with double-stars.

Double-stars had long been known, and were generally recognized. But whether the doubleness were purely accidental, due merely to the fact of two stars happening to lie almost in the same line of sight, as viewed from earth, or whether any real connection existed between the two, no man living could say. Indeed, so long as all stars were regarded as utter fixtures, the question was of no very great interest.

But light dawned as Herschel watched. He found the separate stars in a certain pair to be moving. Each from time to time had slightly, very slightly, changed its place. Then, at least, if all other stars in the universe were fixed, those two were not fixed. He watched on, and gradually he made out that their motions were steady, were systematic, and were connected the one with the other. That was one of the first steps towards breaking down the olden notion, so widely held, of a fixed and unchanging universe.

Another double-star, and yet another, responded to Herschel's intense and careful searching. These other couples, too, were revolving, not separately, but in company; journeying together round one center; bound together apparently by bonds of gravitation, even as our sun and his planets are bound together.

Other stars besides binary stars are found to move with a real and not merely a seeming movement. Viewed carelessly, the stars do indeed appear to remain fixed in changeless groups; fixed even through centuries. But, to exceedingly close watching and accurate measurement, many among them are distinctly *not* fixed; many among them can be actually seen to move.

Of course the observed motions are very small and slow. One may be found to creep over a space as wide as the whole full moon in the course of three hundred or four hundred years; and this is rapid traveling for a star in earth's sky! Another will perhaps cross a space one-tenth or one-twentieth or one-fiftieth of the moon's width in one hundred years.

Such motions had never been carefully noted or examined, until Herschel came to do away with the old received notions of star-fixity. Happily, Herschel was no slave to "received ideas." Like Galileo in earlier times, he wished to "prove all things" personally, anxious only to find out what was the truth. And he found that the stars were *not* fixed! He found that numbers of them were moving. He conjectured that probably all the rest were moving also; that in place of a fixed universe of changeless stars we have a whirling universe of rushing suns.

Herschel could not, of course, watch any one star for a hundred years, much less for several centuries. But in a very few years he could, by exceedingly close measurement, detect sufficient motion to be able to calculate how long it would take a certain star to creep across a space as wide as the full moon.

From step to step he passed on, never weary of his toil. He sought to gauge the Milky Way, and to form some notion of its shape. He noted a general drift of stars to right and to left, which seemed to speak of a possible journey of our sun through space, with all the planets of the Solar System. He flung himself with ardor into the study of star-clusters and of nebulæ. He saw, with an almost prophetic eye, the wondrous picture of a developing universe—of nebulæ growing slowly into suns, and of suns cooling gradually into worlds—so far as to liken the heavens to a piece of ground, containing trees and plants in every separate stage of growth.

And still the search for star parallax went on, so long pursued in vain. No longer base-line than that of the diameter of earth's yearly orbit lay within man's reach. But again and yet again the attempt was made. Instruments were improved, and measurements became ever more delicate; and at length some small success crowned these persistent endeavors. The tiny sounding-line of earth, lowered so often into the mysterious depths of space, did at last "touch bottom."

Herschel died, full of years and honors, in 1822;

and some ten years later three different attempts proved, all to some extent, and almost at the same date, successful. Bessel, however, was actually the first in point of time; and his attempt was upon the double-star, 61 Cygni; not at all a bright star, but only just visible to the naked eye.

There are stars and stars, enough to choose from. The difficulty always was, which to select as a subject for trial with any reasonable prospect of a good result. Some astronomers held that the brightest stars were the most hopeful, since they were probably the nearest; and of course the nearer stars would show parallax more readily, because their parallax would be the greater. But certain very bright stars indeed are now known to be far more distant than certain very dim stars.

Again, some astronomers thought that such stars as could be perceived to move most rapidly in the course of years would be the most hopeful objects to attack, since the more rapid movement might be supposed to mean greater nearness, and so greater ease of measurement. But some stars, seen to move rapidly, are now known to be more distant than others which are seen to move more sluggishly. So neither of these two rules could altogether be depended on; yet both were, on the whole, the best that could be followed, either separately or together.

The star 61 Cygni is not one of the brighter stars, but it is one of those stars which can be seen to travel most quickly across the sky in the course of a century. Therefore it was selected for a trial; and that trial was the first to meet with success.

For 61 Cygni was found to have an apparent parallax; in other words, its tiny seeming journey through the year, caused by our earth's great journey round the sun, could be detected. Small as the star-motion was, it might, through careful measurement, be perceived. And, in consquence, the distance of the star from earth could be measured. Not measured with anything like such exactitude as the measurements of sun-distance, but with enough to give a fair general notion of star-distance.

One success was speedily followed by others. By a few others,—not by many. Among the thousands of stars which can be seen by the naked eye, one here and one there responded faintly to the efforts made. One here and one there was found to stir slightly in the sky, when viewed from earth's summer and winter positions, or from her spring and autumn positions, in her yearly pathway round the sun.

It was a very, very delicate stir on the part of the star. Somewhat like the difference in position which you might see in a penny a few miles off, if you looked at it first out of one window and then out of another window in the same house. You may picture the penny as radiantly bright, shining through pitch darkness; and you may picture yourself as looking at it through a telescope. But even so, the apparent change of position in the penny, viewed thus, would be exceedingly minute, and exceedingly difficult to see.

There are two ways of noting this little seeming movement on the part of a star in the sky.

Either its precise place in the sky may be observed—its exact position, as in a map of the heavens

—or else its place may be noted as compared with another more distant star, near to it in the sky, though really far beyond. Parallax, if visible, would make it alter its precise place in the sky. Parallax, if visible, would bring it nearer to or farther from any other star more distant than itself from earth. The more distant star would have a much smaller parallax—probably so small as to be invisible to us—and so it would do nicely for a "comparison star."

The first of these methods was the first tried; and the second was the first successful.

To find star-distance through star-parallax sounds quite simple, when one thinks of the general principle of it. Just merely the question of a base-line, accurately measured; and of two angles, accurately observed; and of another angle, a good way off, accurately calculated,—all resting on the slight seeming change of position in a certain distant object, watched from two different positions, about 185,000,000 miles apart.

Quite simple, is it not? Only, when one comes to realize that the "change of position" is about equal to the change of position in a penny piece, miles away, looked at from two windows in one house,—then the difficulty grows.

Besides this, one has to remember all the "corrections" necessary, before any true result can be reached.

A penny piece, miles away, seen from two windows, would be difficult enough as a subject for measurement. But, at least, the penny would be at rest; and you yourself would be at rest; and light would pass instantaneously from it to you; and there would be no wobbling and nodding motions of everything around to add to your perplexities.

In the measurement of star-parallax, all these things have to be considered and allowed for; all have to be put out of the question, as it were, before any correct answer is obtained.

The refraction of light must be considered; because that displaces the star, and makes it seem to us to be where it is not. And the aberration of light must be considered; because, in a different way, that does the same thing, making the star seem to take a little journey in the course of the year. And the precession, or forward motion, of the equinoxes, and nutation, or vibratory motion, have both to be separately considered; because they, too, affect the apparent position of every star in the sky.

To watch the star in comparison with another star is easier than merely to note its exact position in the sky; for the other star—the "companion-star"—is equally with itself affected by refraction of light and by aberration of light. But, then, another question comes in seriously: whether or no the companion-star shows any parallax also; since, if it does, that parallax must be carefully calculated and allowed for.

So the measurement of star-distance, even when parallax can be detected, is by no means a light or easy matter. On the contrary, it bristles with difficulties. It is a most complicated operation, needing profound knowledge, accurate observation, trained powers of reasoning and calculation.

Nothing short of what has been termed "the terrific accuracy" of the present day could grapple with the truly tremendous difficulties of this problem of star-distance. It has, however, been grappled with, and grappled with successfully. We now know, not indeed with anything like exactitude, yet with reasonable certainty, the distances of a good many stars, as expressed roughly in round number of "about" so many trillions of miles.* We have at least learned enough to gain some notion of the immeasurable distances of countless other stars, lying far beyond reach of earth's longest measuring-line.

The very thought of such unimaginable depths of space, of suns beyond suns "in endless range," toned down by simple distance to mere quivering specks of light, is well-nigh overwhelming.

---

*The diameter of the moon as seen from the earth is equal to one thousand eight hundred and sixty eight seconds of space. When the parallax of a star is ascertained, it is a simple matter to calculate the distance of the star by the use of the following table. If the star shows a displacement of one second in space, or one eighteen-hundred-and-sixty-eighth part of the width of the moon, that star is distant two hundred and six thousand two hundred and sixty-five times the distance of the earth from the sun.

An	angle	of	1″	is	equivalent	to	206,265	times	93,000,000	of	miles.
"	"	"	0″.9	"	"	"	229,183	"	"	"	"
"	"	"	0″.8	"	"	"	257,830	"	"	"	"
"	"	"	0″.7	"	"	"	294,664	"	"	"	"
"	"	"	0″.6	"	"	"	343,750	"	"	"	"
"	"	"	0″.5	"	"	"	412,530	"	"	"	"
"	"	"	0″.4	"	"	"	515,660	"	"	"	"
"	"	"	0″.3	"	"	"	687,500	"	"	"	"
"	"	"	0″.2	"	"	"	1,031,320	"	"	"	"
"	"	"	0″.1	"	"	"	2,062,650	"	"	"	"
"	"	"	0″.0	"	"	"	Immeasurable.				

The nearest star, Alpha Centauri, shows a parallax of less than one second (0″.75). The farthest star, the distance of which has been ascertained, is 1830 Groombridge, which shows a parallax of only 0″.045, and is distant 4,583,000 times earth's distance from the sun, or 426 trillions of miles.

CHAPTER XXVIII.

THE MILKY WAY.

THE MILKY WAY forms a soft band of light round the whole heavens. In the Southern Hemisphere, as in the Northern Hemisphere, it is to be seen. In some parts the band narrows; in some parts it widens. Here it divides into two branches; there we find dark spaces in its midst. One such space in the south is so black and almost starless as to have been named the Coal-sack. All along, over the background of soft, dim light, lies a scattering of brighter stars shining on its surface. Much interest and curiosity have long been felt about this mysterious Milky Way. That it consists of innumerable suns, and that our sun is one among them, has been believed for a considerable time. But other questions arise. How many stars does the Milky Way contain? What is its shape? How far does it reach?

No harm in asking the questions, only we have to be satisfied in astronomy to ask many questions which can not yet receive answers. No harm for man to learn that the utmost reach of his intellect must fall short in any attempt to sound the depths of God's universe, even as the arm of a child would fall short in seeking to sound the depths of the ocean over the side of a little boat. For the attempt has been made to sound the depths of our star-galaxy out of this little earth-boat.

The idea first occurred to the great Herschel, as we have already mentioned, and a grand idea it was—only a hopeless one. He turned his powerful telescope north, south, east, west. He counted the stars visible at one time in this, in that, in the other directions. He found a marked difference in the numbers. The portion of sky seen through his telescope was about one quarter the size of that covered by the moon. Sometimes he could merely perceive two or three bright points on a black background. At other times the field of his telescope was crowded. In the fuller portions of the Milky Way, he had four or five hundred stars under view at once. In one place he saw about one hundred and sixteen thousand stars pass before him in a single quarter of an hour.

Herschel took it for granted that the stars of the Milky Way, uncountable in numbers, are, as a rule, much the same in size—so that brightest stars would, as a rule, be nearest, and dimmest stars would, as a rule, be farthest off. Where he found stars clustering thickly, beyond his power to penetrate, he believed that the Milky Way reached very far in that direction. Where he found black space, unlighted by stars or lighted by few stars, he decided that he had found the borders of our galaxy in that direction.

Following these rules which he had laid down, he made a sort of rough sketch of what he supposed might be the shape of the Milky Way. He thought it was somewhat flat, extending to a good distance breadthways and a much greater distance lengthways, and he placed our sun not far from the middle. This imagined shape of the Milky Way is called "The

Cloven-disk Theory." To explain the appearance of the Milky Way in the sky, Herschel supposed it to be cloven or split through half of its length, with a black space between the two split parts.

It seems that Herschel did not hold strongly to this idea in later years, and doubts are now felt whether the rules on which he formed it have sufficient foundation.

For how do we know that the stars of the Milky Way are, as a rule, much the same in size? Certainly the planets of the Solar System are very far from being uniform, and the few stars whose weight can with any certainty be measured, seem to vary considerably. There is a great difference also between the large and small suns in many of the double-stars!

Again, how do we know that the bright stars are, as a rule, the nearest, and the dim stars the farthest off? Here, also, late discoveries make us doubtful. Look at Sirius and 61 Cygni—Sirius the most radiant star in the heavens, and 61 Cygni almost invisible to the naked eye. According to this rule, 61 Cygni ought to lie at an enormous distance beyond Sirius. Yet in actual fact, Sirius is the farthest away of the two.

The illustrious Herschel believed that he had penetrated, on one occasion, into the star-cluster on the sword-hand in the constellation of Perseus until he found himself among siderial depths, from which the light could not have reached him in less than four thousand years.

The distance of those stars had not, and has not, been mathematically measured. Herschel judged of it by their dimness, by the strong power needed to make them visible, and by the rules which he had adopted as most likely true.

Again, when Herschel found black spaces in the heavens almost void of stars, and believed that he had reached the outside borders of the Milky Way, he may have been in the right, or he may have been

A CLUSTER OF STARS IN PERSEUS.

mistaken. The limit might lie there, or thousands more of small stars might extend in that very direction, too far off for their little glimmer to be seen through the most powerful telescope.

If this latter idea about the Milky Way being formed of a great many brilliant suns, and of vast numbers of lesser suns also, be true, astronomers will, in time, be able to prove its truth. For in that case, many faint telescopic stars being much nearer to us than bright stars of the greater magnitudes, it will be found possible to measure their distance. The journey of our earth round the sun must cause a seeming change in their position between summer and winter.

The theory also that some of the nebulæ are other outlying Milky Ways, or galaxies of stars, separated by tremendous distances from our own, is interesting, and was long held as almost certain, yet we have no distinct proof either one way or the other. Many of the nebulæ may be such gatherings of countless stars outside our own, or every nebula visible may be actually part and parcel of our galaxy.

Much attention has of late been paid to the arrangement of stars in the sky. The more the matter is looked into, the more plainly it is seen that stars are neither regular in size nor regular in distribution. They are not merely scattered carelessly, as it were, here, there, and anywhere, but certain laws and plans of arrangement seem to have been followed which astronomers are only now beginning dimly to perceive.

Stars are not flung broadcast through the heavens, each one alone and independent of the rest. They are placed often, as we have already seen, in pairs, in triplets, in quartets, in clusters. Also, the great masses of them in the heavens seem to be more or less arranged in streams, and sprays, and spirals. So remarkable are the numbers and forms of many of these streams that the idea has been suggested, with regard to the Milky Way, whether it also may not be a vast stream of stars, like a mighty river, collecting into itself hundreds of lesser streams.

The contemplation of the heavens affords no spectacle so grand and so eloquent as that of a cluster of stars. Most of them lie at such a distance that the most powerful telescopes still show them to us like star-dust. "Their distance from us is such that they are beyond, not only all our means of measurement," says Newcomb, "but beyond all our powers of estimation. Minute as they appear, there is nothing that we know of to prevent our supposing each of them to be the center of a group of planets as extensive as our own, and each planet to be as full of inhabitants as this one. We may thus think of them as little colonies on the outskirts of creation itself, and as we see all the suns which give them light condensed into one little speck, we might be led to think of the inhabitants of the various systems as holding intercourse with each other. Yet, were we transported to one of these distant clusters, and stationed on a planet circling one of the suns which compose it, instead of finding the neighboring suns in close proximity, we should see a firmament of stars around us, such as we see from the earth. Probably it would be a brighter firmament, in which so many stars would glow, with more than the splendor of Sirius, as to make the night far brighter than ours; but the inhabitants of the neighboring worlds would as completely elude telescopic vision as the inhabitants of Mars do here. Consequently, to the inhabitants of every planet in the cluster, the question of the plurality of worlds might be as insolvable as it is to us."

These are clusters of stars of regular form in which attraction appears to mark its secular stamp. Our mind, accustomed to order in the cosmos, anxious for harmony in the organization of things, is satisfied with these agglomerations of suns, with these distant universes, which realize in their entirety an aspect approaching the spherical form. More extraordinary, more marvelous still, are the clusters of stars which appear organized in spirals.

In considering stars of the first six magnitudes only—stars visible to the naked eye—a somewhat larger number is found in the Southern Hemisphere than in the Northern Hemisphere. In both hemispheres there are regions densely crowded with stars, and regions by comparison almost empty.

It has been long questioned whether the number of bright stars is or is not greater in the Milky Way than in other parts of the sky. Careful calculations have at length been made. It appears that the whole of the Milky Way—that zone of soft light passing round the earth—covers, if we leave out the Coal-sack and other such gaps, between one-tenth and one-eleventh of the whole heavens.

The entire number of naked-eye stars, or stars of the first six magnitudes, does not exceed six thousand; and of these, eleven hundred and fifteen lie scattered along the bed of the Milky Way stream. If the brighter stars were scattered over all the sky as thickly as throughout the Milky Way, their number would amount to twelve thousand instead of only six thousand. This shows us that the higher-magnitude stars really are collected along the Milky Way in greater numbers than elsewhere, and is an argument used by those who believe the Milky Way to be a mighty stream of streams of stars.

In the dark spaces of the Milky Way, on the contrary, bright stars are so few that if they were scattered in the same manner over all the sky, their present number of six thousand would come down to twelve hundred and forty. This would be a serious loss.

We know in the sky 1,034 clusters of stars and more than 11,000 nebulæ. The former are composed of associated stars; the latter may be divided into two classes: First, nebulæ which the ever-increasing progress of optics will one day resolve into stars, or which in any case are composed of stars, although their distances may be too great to enable us to prove it; second, nebulæ properly so called, of which spectrum analysis demonstrates their gaseous constitution. Here is an instructive fact. The clusters of stars present the same general distribution as the telescopic stars—they are more numerous in the plane of the Milky Way, while it is the contrary which is presented by the nebulæ properly so called; they are rare, thinly spread in the Milky Way, and thickly scattered to the north as well as to the south of this zone up to its poles. The constitution of the Milky Way—not nebulous, but stellar—is a very significant fact. The nebulæ properly so called are distributed, in a sense, contrary to the stars, being more numerous towards the poles of the Milky Way and in regions poor in stars, as if they had absorbed the matter of which the stars are formed.

CHAPTER XXIX.

A WHIRLING UNIVERSE.

NO REST, no quiet, no repose, in that great universe, which to our dim eyesight looks so fixed and still, but one perpetual rush of moving suns and worlds. For every star has its own particular motion, every sun is pressing forward in its own appointed path. And among the myriads of stars—bright, blazing furnaces of white or golden, red, blue, or green flame—sweeping with steady rush through space, our sun also hastens onwards.

The rate of his speed is not very certain, but it is generally believed to be about one hundred and fifty million miles each year. Possibly he moves in reality much faster.

When I speak of the sun's movement, it must of course be understood that the earth and planets all travel with him, much as a great steamer on the sea might drag in his wake a number of little boats. From one of the little boats you could judge of the steamer's motion quite as well as if you were on the steamer itself. Astronomers can only judge of the sun's motion by watching the seeming backward drift of stars to the right and left of him; and the watching can be as well accomplished from earth as from the sun himself.

After all, this mode of judging is, and must be, very uncertain. Among the millions of stars visible, we only know the real distances of about twenty-five; and every star has its own real motion, which has to be separated from the apparent change of position caused by the sun's advance.

It seems now pretty clear that the sun's course is directed towards a certain point in the constellation Hercules. If the sun's path were straight, he might be expected by and by, after long ages, to enter that constellation. But if orbits of suns, like orbits of plan-

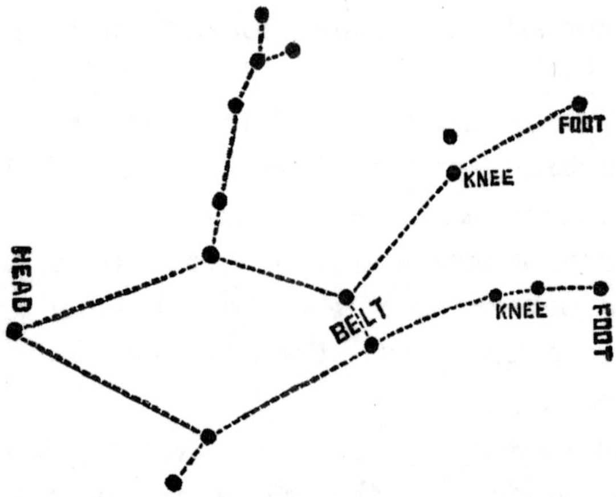

HERCULES

ets, are ellipses, he will curve away sideways long before he reaches Hercules.

One German astronomer thought he had found the center of the sun's orbit. He believed the sun and the stars of the Milky Way to be traveling round the chief star in the Pleiades, Alcyone. This is not impossible; but it is now felt that much stronger proof will be required before the idea can be accepted.

In the last chapter mention was made of star-streams as a late discovery. Though a discovery still in its infancy, it is one of no small importance. Briefly stated, the old theory as to the plan of the Milky Way was as follows: Our sun was a single star among millions of stars forming the galaxy, some comparatively near, some lying at distances past human powers of calculation, and all formed upon much the same model as to size and brightness. Where the band of milky light showed, stars were believed to extend in countless thousands to measureless distances. Where dark spaces showed, it was believed that we looked beyond the limits of our universe into black space. Stars scattered in other parts of the sky were supposed generally to be outlying members of the same great Milky Way. Many of the nebulæ and star-clusters were believed to be vast and distant gatherings of stars, like the Milky Way itself, but separated by unutterably wide reaches of space. Some of these views may yet be found to contain truth, though at the present moment a different theory is afloat.

It is still thought probable that our sun is one among many millions of suns, forming a vast system or collection of stars, called by some a universe. It is also thought possible that other such mighty collections of stars may exist outside and separate from our own at immense distances. It is thought not impossible or improbable that some among the nebulæ may be such far-off galaxies of stars; though, on the other hand, it is felt that every star-cluster and nebula within reach of man's sight may form a part of our own "universe."

According to this view of the question, the Milky Way, instead of being an enormous universe of countless suns reaching to incalculable distances, may rather be a vast and mighty star-stream, consisting of hundreds of brilliant leading suns, intermixed with thousands or even millions of lesser shining orbs. If this be the true view, the lesser suns would often be nearer than the greater suns, although more dim; and the Milky Way would not be itself a universe, though a very wonderful and beautiful portion of our universe.

Which of these two different theories or opinions contains the most truth remains to be found out. But respecting the arrangement of stars into streams, interesting facts have lately been discovered. We certainly see in the Solar System a tendency of heavenly bodies to travel in the same lines and in companies. Not to speak of Jupiter and his moons, or Saturn and his moons, we see it more remarkably in the hundreds of asteroids pursuing one path, the millions of meteorites whirling in herds. Would there be anything startling in the same tendency appearing on a mightier scale? Should we be greatly astonished to find streams of stars, as well as streams of planets?

For such, indeed, appears to be the case. Separated by abysses of space, brother suns are plainly to be seen journeying side by side through the heavens, towards the same goal.

The question of star-drift is too complex and difficult to be gone into closely in a book of this kind. One example, however, may be given. Almost everybody knows by sight the constellation of the Great Bear—the seven principal stars of which are called also Charles's Wain, a corruption of the old Gothic *Karl Wagen*, the churl's or peasant's wagon. It is also called the Great Dipper. Four bright stars form a rough sort of oblong, and from one of the corners three more bright stars stretch away in a curve, representing the Bear's tail. Many smaller stars are intermixed.

These seven bright stars have always been bound together in men's minds, as if they belonged to one

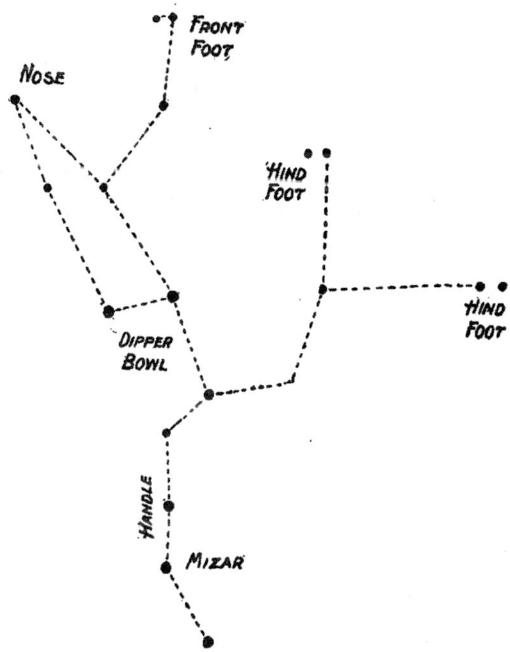

THE GREAT BEAR.

another. But who, through the centuries past, since aught was known of the real distances of the stars from ourselves and from one another, ever supposed that any among the seven were *really* connected together?

One of these seven stars, the middle one in the tail, has a tiny companion-star, close to it, visible to the naked eye. For a good while it was uncertain whether the two were a "real double," or only a seeming double. In time it became clear that the two did actually belong to one another.

Mizar is the name of the chief star, and Alcor of the companion. Alcor is believed to be about three thousand times as far away from Mizar, as our earth from the sun. Now, if this be the width of space between those two bright points, lying seemingly so close together as almost to look to the naked eye like one, what must be the distance between the seven leading stars of the Great Bear, separated by broad sky-spaces? Who could imagine that one of these suns had aught to do with the rest?

Yet among other amazing discoveries of late years it has been found by means of the new instrument, the spectroscope, that *five* out of these seven suns are traveling the same journey, with the same speed. Two of the seven appear to be moving in another direction. but three of the body-stars and two of the tail-stars are hastening in the direction away from us, all in the same line of march, all rushing through space at the rate of twenty miles each second. Some smaller stars close to them are also moving in the same path. Is not this wonderful? We see here a vast system of suns, all moving towards one goal, and each probably bearing with him his own family of worlds.

Many such streams have been noticed, and many more will doubtless be found. For aught we know, our own sun may be one among such a company of brother-suns, traveling in company.

It is difficult to give any clear idea of the immensity of the universe—even of that portion of the universe which lies within reach of our most powerful telescopes. How far beyond such limits it may reach, we lose ourselves in imagining.

Earlier in the book we have supposed possible models of the Solar System, bringing down the sun and worlds to a small size, yet keeping due proportions. What if we were to attempt to make a reduced model of the universe; that is of just so much of it as comes within our ken?

Suppose a man were to set himself to form such a model, including every star which has ever been seen. Let him have one tiny ball for the sun, and another tiny ball for Alpha Centauri; and let him, as a beginning, set the two *one yard apart.* That single yard represents ninety-three millions of miles, two hundred and seventy-five thousand times repeated. Then let him arrange countless multitudes of other tiny balls, at due distances—some five times, ten times, twenty times, fifty times, as far away from the sun as Alpha Centauri.

It is said that the known universe, made upon a model of these proportions, would be many miles in length and breadth. But the model would appear fixed as marble. The sizes and distances of the stars being so enormously reduced, their rates of motion would be lessened in proportion. Long intervals of time would need to pass before the faintest motion in one of the millions of tiny balls could become visible to a human eye.

CHAPTER XXX.

READING THE LIGHT.

SEVERAL times, in the course of this book, mention has been made of a wonderful new instrument called the spectroscope. We now proceed to give a brief account of the origin and achievements of this newest of scientific appliances. The first step towards its

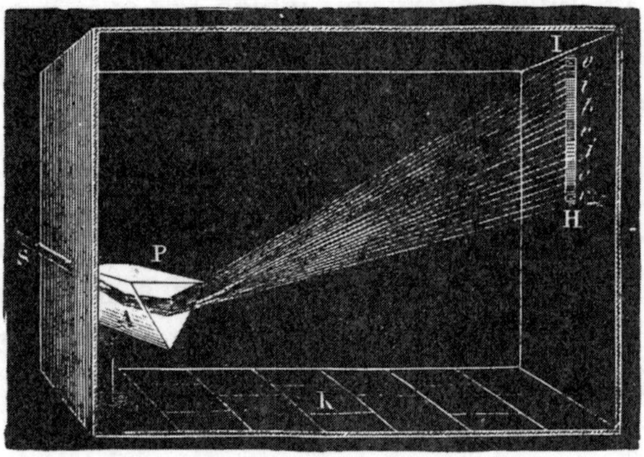

THE PRISM AND SPECTRUM.

formation was made by Sir Isaac Newton, when he discovered the power of the prism to decompose light. This consists in the fact that a ray of light, after passing through a transparent prism, becomes expanded into an elongated spectrum, no longer white, but presenting an invariable succession of colors from red to violet. These are called the seven primary colors; namely, red, orange, yellow, green, blue, indigo, and violet. The rainbow is a familiar illustration of this spectrum with its various colors; and the raindrops are the prisms which reflect and decompose the light.

Optical science was long satisfied with this glance into the interior constitution of light, occupying itself with the phenomena of the prismatic colors, and theorizing on the nature of white light. In 1802, Dr. W. H. Wollaston, in closely examining a spectrum, found it to be crossed by at least four fine dark lines. It is only when an extremely narrow slit is employed

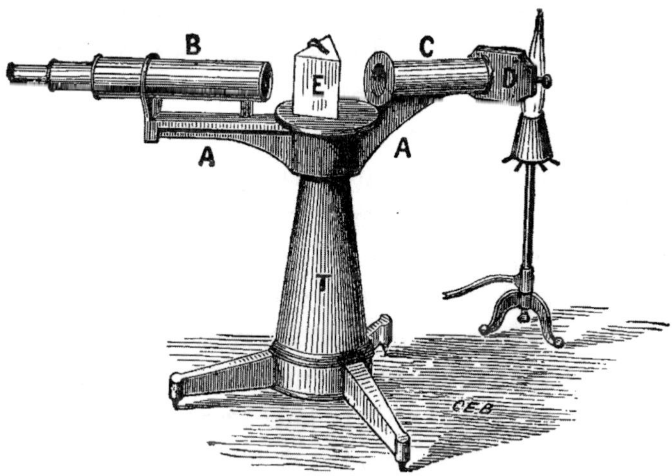

THE SPECTROSCOPE.

in admitting the sunlight that they become visible. Dr. Wollaston, supposing them to be merely "natural boundaries" of the different color-bands, inquired no further; and there for a while the matter rested. Not many years later, in 1815, the matter was taken up by a German optician and scientist of Munich, Joseph von Fraunhofer. Applying more delicate means of observation, he was surprised to find very numerous dark lines crossing the spectrum.

With patience he went into the question, using the

telescope as well as a very narrow slit, and soon he discovered that the dark lines were to be numbered, not by units or by tens, but by hundreds—or, as we now know, by thousands. Some were in the red, some were in the violet, some were in the intermediate bands; but each one had, and has, its own invariable position on the solar spectrum. For, be it understood, these dark lines are constant, not variable. Where a line is seen, there it remains. Whenever a ray of sunlight is properly examined, with slit and with prism, that line will be found always occupying precisely the same spot in the spectrum.

Some of the chief and more distinct lines were named by Fraunhofer after certain letters of the alphabet, and by those letter-names they are still known. He began with A in the red and went on to H in the violet. Fraunhofer made a great many experiments connected with these mysterious lines, anxious to discover, if possible, their meaning. For although he now saw the lines, which had scarcely so much as been seen before, he could not understand them—he could not read what they said. They spoke to him, indeed, about the sun; but they spoke in a foreign language, the key to which he did not possess. He tried making use of prisms of different materials, thinking that perhaps the lines might be due to something in the nature of the prism employed. But let the prism be what it might, he found the lines still there. Then he examined the light which shines from bright clouds, instead of capturing a ray direct from the sun. And he found the lines still there. For cloudlight is merely reflected sunlight.

Then he examined the light of the moon, to see if perchance the spectrum might be clear of breaks. And he found the lines still there. For moonlight is only reflected sunlight. Next he set himself to examine the light which travels to us from some of the planets, imagining that a different result might follow. And he found the lines still there. For planet-light again is no more than reflected sunlight.

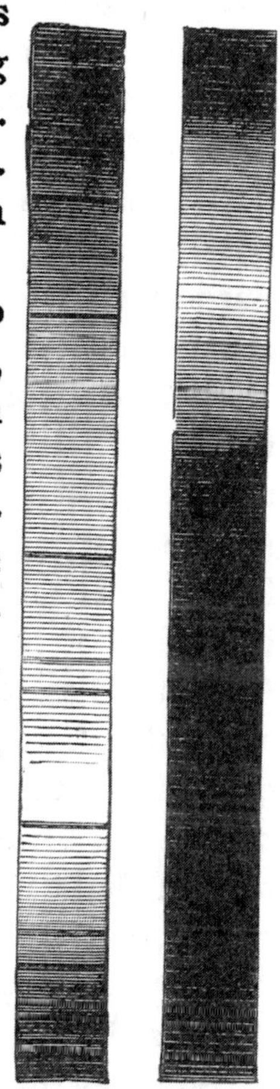

SPECTRA, SHOWING THE DARK LINES.

Lastly, he turned his attention to some of the brighter stars, examining, one by one, the ray which came from each. And, behold! he found the lines *not* there. For starlight is not reflected sunlight. That is to say, the identical lines which distinguish sunlight were not there. Each star had a spectrum as the sun has a spectrum, and each star-spectrum was crossed by faint dark lines, more or less in number. But the spectra of the stars differed from the spectrum of the sun. Each particular star had its own particular spectrum of light, different from that of the sun, and different from that of every other star. For now, Fraunhofer was examining, not sunlight, but starlight; not the light of our sun, either direct from himself or reflected from some other body, such as planet or moon or cloud, but the light of other suns very far distant, each one varying to some extent from the rest in its make. The fact of the stars showing numerous sets of black lines, all unlike those of our sun, showed conclusively that those lines could not possibly be due to anything in our earthly atmosphere. Sunlight and starlight travel equally through the air, and are equally affected by it. If our atmosphere were the cause of the black lines in sunlight, it would cause the *same* lines in starlight. But the sun and each individual star has its own individual lines, quite irrespective of changeful states of the air.

So, also, the light from any metal sufficiently heated will give a spectrum, just as sunlight gives a spectrum, under the needful conditions. That is to say, there must be slit and prism, or slit and diffraction-grating, for the light to pass through. But the kind of spectrum is by no means always the same.

Putting aside for a few minutes the thought of sunlight and starlight, let us look at the kind of rays or beams which are given forth by heated earthly substances. Any very much heated substance sends forth its light in rays or beams, and any such rays or beams may be passed through a prism, and broken up or "analyzed," just as easily as sunlight may be "analyzed."

Suppose that we have a solid substance first—a piece of iron or of steel wire. If it is heated so far as to give out, not only heat, but also light, and if that light is made to travel through the slit and prism of a spectroscope, the ray will then be broken up into its sub-rays. They, like the sub-rays of a sunbeam, will form a continuous row of soft color-bands, one melting into another. This is the characteristic spectrum of the light which is given forth by a burning or glowing *solid*.

Next, suppose we take a liquid—some molten iron or some molten glass, for instance. If you have ever been to a great plate-glass manufactory, like that at St. Helens, not far from Manchester, you will have seen streams of liquid glass pouring about, carried to and fro in huge caldrons, bright with a living light of fire from its intensity of heat. If a ray of *that* light had been passed through slit and prism, what do you think would have been the result? A continuous spectrum once more, the same as with the glowing iron or steel. The light-ray from a heated and radiant *liquid*, when broken up by a prism, lies in soft bands of color, side by side.

Both of them a good deal like the solar spectrum, you will say. Only here are no mysterious dark lines crossing the bright bands of color. But how about gases? Suppose we have a substance in the state of gas or vapor, as almost every known substance might be under the requisite conditions, and suppose that substance to be heated to a glowing brilliance. Then let its light be passed through the slit and prism of a spectroscope. What result shall we find this time? Entirely different from anything seen before. Instead of soft, continuous bands of color, there are *bright lines*, well separated and sharply defined.

How many bright lines? Ah, that depends upon which particular gas is having its ray analyzed. Try sodium first—one of the commonest of earthly substances. Enormous quantities of it are distributed broadcast in earth and air and water. More than two-thirds of the surface of our globe lies under an enfolding vesture of water saturated with salt, which is a compound of sodium. That is to say, sodium enters largely into the make of salt. Every breeze which sweeps over the ocean carries salt inland, to float through the atmosphere. Sir H. E. Roscoe writes: "There is not a speck of dust, or a mote seen dancing in the sunbeam, which does not contain chloride of sodium"—otherwise salt. The very air which surrounds us is full of compounds of sodium, and we can not breathe without taking some of it into our bodies. Sodium is an "elementary substance." By which I mean that it is one of a number of substances called by us "simple," because chemists have never yet succeeded in breaking up those substances, by any means at their command, into other and different materials.

Iron is, so far as we know, a simple substance. It is found as iron in the earth. No chemist has ever been able to make iron by combining other materials together. No chemist has ever managed to separate iron into other materials unlike itself. Iron it is, and iron it remains, whether as a solid, as a liquid, or as a gas. Gold is another simple substance, and so is silver. Sodium is another. All these we know best in the solid form. Mercury, another simple substance, we know best as a liquid.

Water is not a simple substance; for it can be separated into two different gases. Glass is not a simple substance; for it is manufactured out of other substances combined together.

We can speak quite positively as to such substances as are not simple; but with regard to so-called "elements," we may only venture to assert that, thus far, nobody has succeeded in breaking them up. Therefore, at least for the present, they are to us "elementary."

All the simple substances, and very many of the combinations of them, though often known to us only in the solid form, may, under particular conditions, be rendered liquid, and even gaseous. Every metal may be either in the solid form, or the liquid form, or the vapor form. Iron, as we commonly see it, is solid—in other words, it is frozen, like ice. Just as increase of warmth will turn ice into water, so a certain amount of heat will make solid iron become liquid iron. And just as yet greater warmth will turn water into steam, so a very much increased amount of heat will turn liquid iron into vapor of iron. A little heat will do for ice what very great heat will do for iron.

Whether iron and other metals exist in the sun, as on earth, in a hard and solid form, it is impossible to say. It is only in the form of gas that man can become aware of their presence at that distance. The intense, glowing, furnace-heat of the sun causes many metals to be present in large quantities in the sun's atmosphere in the form of vapor.

Not only have we learned about some of the metals in the sun, but this strange spectrum analysis has taught us about some of the metals and gases in the stars as well. It is found that in Sirius, sodium and magnesium, iron and hydrogen, exist. In Vega and Pollux there are sodium, magnesium, and iron. In

Aldebaran, these substances and many others, including mercury, seem to abound. These are merely a few examples among many stars, each being in some degree different from the rest.

But how can we know all this? How could the wildest guessing reveal to us the fact of iron in the sun, not to speak of the stars? We know it by means of the spectrum analysis—or, as we may say, by means of the spectroscope. This instrument may be looked upon as the twin-sister to the telescope. The telescope gathers together the scattered rays of light into a small spot or focus. The spectroscope tears up these rays of light into ribbons, sorts them, sifts them, and enables us to read in them hidden meanings.

When a ray of light reaches us from the sun, that ray is *white;* but in the white ray there are bright colors concealed. Newton was the first to discover that a ray of white light is really a bundle of colored rays, so mixed up together as to appear white. If a ray of sunlight is allowed to pass through a small round hole in a wall, it will fall upon the opposite wall in a small round patch of white light. But if a *prism*—a piece of glass cut in a particular shape—is put in the path of the ray, it has power to do two curious things. First, it bends the ray out of a straight course, causing the light to fall upon a different part of the wall. Secondly, it breaks up or divides the ray of white light into the several rays of colored light of which the white ray is really composed. This breaking up, or dividing, is called "analyzing."

If in place of a round hole the ray of light is made

to pass through a very narrow slit, it is proved that the bright bands of color do not overlap. Instead of this, dark lines, or gaps, show here and there. Now, these lines or gaps are always to be seen in the spectrum or image of bright colors formed by a broken-up ray of *sunlight*. There is always a certain number of dark lines in each colored band—some near together, some far apart; here one or two, there a great many. Where a simple ray of sunlight is concerned, the exact arrangement of the lines never changes.

When the stars were examined—when the rays of light coming from various stars were split up and analyzed—it was found that they too, like the sun, gave a spectrum of bright colors with dark lines. But the lines were different in number and different in arrangement from the sun's lines. Each star has his own particular number and his own particular arrangement, and that arrangement and number do not change.

If a white-hot metal is burnt, and the light of it as it burns is allowed to pass through a prism, a row of bright colors appears as in the sun's spectrum, only there are no dark lines. If a gas is burnt, and the light is allowed to pass through a prism, no bright color-bands appear, and no dark lines either; but instead of this, there are *bright lines*. Each gas or vapor has its own number of lines and its own arrangement. Sodium shows two bright lines, side by side. Iron shows sixty bright lines, arranged in a particular way.

Now you see how a row of bright colors without bands of color may appear. But what about color-bands and dark lines together? That discovery came latest. It was found that if a white-hot metal were burned, and if its light were allowed before touching the prism to shine through the flame of a burning gas, *then* there were dark lines showing in the colored bands. These dark lines changed in position and number and arrangement with each different kind of gas, just as the bright lines changed if the gases were burned alone. If the light of the burning metal passed through a flame colored with gas of sodium, two dark lines showed on one part of the spectrum; but if it passed through a flame colored with vapor of iron, sixty dark lines showed on another part of the spectrum.

So now, by means of this spectrum analysis, we know with all but certainty that the sun and stars are solid, burning bodies, sending their light through burning, gas-laden atmospheres. By examining the little black lines which appear in the spectrum of one or another, it is possible to say the names of many metals existing as gas in those far-off heavenly bodies. Is not this a wonderful way of reading light?

The split-up rays tell us much more than the kinds of metals in different stars. When a nebula is examined, and is found to give no spectrum of bright bands and dark lines, but only a certain number of bright lines, we know it to be formed of gas, unlike stars and other nebulæ. Also, it is by means of the spectroscope that so much has lately been discovered about the motions and speed of the stars coming towards or going from us.

CHAPTER XXXI.

STELLAR PHOTOGRAPHY.

PHOTOGRAPHY is not only a useful handmaid to astronomy as a whole, but also it is so, peculiarly, to that division of astronomy with which we are now chiefly concerned—to spectroscopy. A few pages may therefore with advantage be given to the subject of celestial photography.

Photography is, indeed, the greatest possible help to the astronomer. It pictures for him those stars which his eyes can see but dimly, or even can not see at all; it paints for him those light-rays of which he would obtain but a passing glance, and which he could not accurately remember in all their details; it maps out for him the wide heavens, which he, unaided, could never do with anything like equal completeness by eye and hand alone.

Only recently a vast photographic map of the whole sky was undertaken. About eighteen different observatories, in divers parts of the world, divided the task among them, stars down to the sixteenth magnitude being most carefully registered in a complete series of something like fifteen hundred separate photographs. The whole result, when finally completed, will be a grand achievement of the present century. Each individual star, in the entire heavens all around our earth, from the first to the sixteenth magnitudes, will have its exact position in the sky accurately known, and the smallest change in the position of any one of those stars may then be detected.

Even in the delicate and abstruse operation earlier described, the measurement of star-distances through annual parallax, photography again steps in. Dr. Pritchard, Savilian professor of Astronomy at Oxford, pressed photography into the service of this task also. Measurements for parallax were made under his direction upon photographic plates—a work of no small interest. Here, as elsewhere, peculiar advantages belong to the photographic method when it can be followed. Its records are lasting, the limited number of hours which are fully suitable for direct astronomical work may be employed in obtaining those records, and in broad daylight the examination of them can be carried on.

One very curious use is made of star-photographs, more especially of the photographs of unseen stars—that is, of stars too distant or too dim to be detected by the eye. These photographs, when taken, may be afterwards looked into further with a microscope. So, first, the far-off, invisible suns of the universe are photographed on a prepared plate, with the help of a powerful telescope, this being needful to secure sufficient starlight; and then the tiny picture of those suns is examined more closely with the help of a powerful microscope.

Spectroscopy has much to say to us. It tells us about the positions of the different stars. It tells us about the structure of the stars. It tells us about the various classes to which the various stars belong. And also it tells us about the motions of the stars—not mere apparent motions, caused by movements of our own earth, but true onward journeyings of the stars themselves through the depths of space.

For by means of photography we do not obtain simple pictures of the stars themselves *only*, but pictures also of the *spectra* of the stars. An instrument is made uniting the spectroscope with the photographic apparatus, and this is called a spectrograph. By its means the disintegrated or broken-up star-ray is photographed in its broken-up condition, so that an exact picture is obtained of the bands and lines characteristic of any particular star.

No easy matter, as may be imagined, are these spectroscopic observations and these spectroscopic photographings of the stars. To bring the image of a star exactly opposite a slender opening or slit, perhaps only about the *three-hundredth part of an inch in width*, and to keep it there, is a task which might well be looked upon as practically impossible.

No sooner is a star found in the field of a telescope than it vanishes again. As the astronomer gazes, the ground beneath him is ever whirling onward, leaving the star behind; and although clockwork apparatus in all observatories of any importance is made to counteract this motion of the earth, and to keep the heavenly object, whether star or moon, or comet, within view by following its apparent motion, yet, as can easily be imagined, to follow thus the seeming motion of a dim star through an opening so minute, demands exceeding care and delicacy of adjustment.

It is not indeed, for purposes of analysis, *always* needful to pass the light of a star through a narrow slit. In the case of a nearer body, such as the sun or moon or one of the planets, light flows from all parts of the body, one ray crossing another; and for the examination of such light, a slit is imperatively needed, all side-rays having to be cut off. But the whole of the brightest star in the sky is only one point of light to us, and a slitless telescope may be used with no confused results. Where, however, direct comparison is required, the star-lines being made to appear, side by side or above and below, with the solar spectrum, or with the lines of earthly metals, then the slit becomes unavoidable.

By such comparison of the two, side by side—the light from an intensely-heated earthly substance and the light from a star, the rays of both being broken up in the spectroscope—the oneness or difference of lines in the rays can be easily made out.

One main difficulty in such observations arises from the diminution of starlight, caused by its passage through a prism. If a rope of a dozen strands is untwisted, each of those strands is far weaker than the whole rope was; and each strand of color in the twisted rope of light is of necessity much more feeble, seen by itself, than the whole white ray seen in one. Another difficulty in our country generally is the climate, which gives so few days or hours in the year for effective work. Yet the full amount accomplished by seizing upon every possible opportunity is, in the aggregate, astonishing. A third and very pressing difficulty attendant upon examination of star-spectra is caused by the incessant motion of our air, through which, as through a veil, all observations have to be made. The astronomer can never get away from the atmosphere; and unless the air be very still—that is to say, as still as it ever can remain—the spectrum-lines are so uncertainly seen as to make satisfactory results impossible. Dr. Huggins has sometimes passed hours in the examination of a single line, unable to determine whether or no it precisely coincided with the comparison-line of some earthly substance. In *this* matter, no leaping at conclusions is admissible.

The photography of stars would be easy enough if one could just expose a plate to the shining of the star, and there leave it to be impressed—there leave the star-ray to sketch slowly its own image. But this is hardly possible. The unceasing motion of the earth, causing the star perpetually to pass away from the telescopic field, and the exceeding narrowness of the slit opposite to which the star has to be kept in a stationary position, make the most accurate adjustment needful.

Clockwork alone can not be trusted. If it could, the star and the photographic apparatus might be comfortably "fixed," and left to do their own work. Instead of which, while a photograph is proceeding, it is desirable that an observer should sit gazing patiently at the telescope-tube, where the image of the star is seen, ready at any moment to correct by a touch the slightest irregularity in the clockwork motion of the telescope, and so to prevent a blurred and spoilt reproduction of the star, or of its spectrum.

One hour, two hours, three hours, at a stretch, this unceasing watchfulness may have to be kept up, and no small amount of enthusiasm in the cause of science is requisite for so monotonous and wearying a vigil.

As noted earlier, it has been found possible, if the photograph of a star or nebula is not completed in one night, to renew the work and carry it on the next night, or even for many nights in succession. This is especially practicable in spectroscopic photography.

From four to five hours, sometimes from eight to ten hours, may be needful before the clear image of the star or of its spectrum appears—a dim little star probably to us, yet perhaps in reality a splendid sun, shedding warmth and light and life upon any number of such worlds as ours.

Eight or ten hours of photography at one stretch with a star are impossible; for the stars, ever seemingly on the move, do not remain long enough in a good position. For three to six hours, a telescope may be made, by means of its clockwork machinery, to keep a star steadily in sight, and all that while the photograph is progressing. If further exposure is needed, the process has to be resumed the next night.

The more one considers the matter, the more plainly one perceives how enormously our powers of sky-observation are increased by photography. It is not only that one photographer, with his apparatus, may accomplish in a single night the work of many astronomers who have to depend upon the power of the eye alone. It is not only that, with the help of photography, as much can now be done in a lifetime as formerly must have occupied many generations. It is not only that the photograph, once taken, remains a permanent possession, instead of a record, more or less imperfect, in which otherwise the astronomer would have to trust.

It is not even only that in the photograph details come out which could not be detected by the eye, and that stars are actually brought to our knowledge which no man has ever seen, which perhaps no man ever will see from this earth with the assistance of the most powerful telescope. For the weak shining, which can by no possibility make itself felt by the retina of a man's eye, *can* slowly impress its picture on the photographic plate. Hundreds of stars, thousands of stars, utterly invisible to man, have had their photographs taken as truly as you have had your photograph taken, and by the same process, only it has been a longer business.

But in addition to all this, we see reproduced upon the plate those ultra-violet and infra-red portions of the spectrum of light which but for the handmaid, photography, would still be to us as things which have no existence. And by means of photography we can observe and study in those same unseen portions of the spectra, when looking into a ray from sun or star, the innumerable dark lines, every one of which has its own tale to tell. To these tales we must have remained blind and deaf, but for photographic aid to our limited powers.

Look at some dim star in the sky, and try how long you can gaze without blinking. You will very soon find that you have done your best, and that to gaze longer only means a sense of fatigue in your eyes, a growing dimness in the star. How different with the photographic plate! There, no exertion is wasted, no weariness is felt. Faint though the light may be, which travels earthward and falls upon the plate, it is all collected, all used.

It is easy to photograph the sun, and this has been done. When we come to the moon the operation is more difficult on account of the feeble intensity of its light, and the difference of tint of the various parts of its surface. Skill and perseverance have, however, surmounted the greatest difficulties, and now we have photographs of the moon enlarged to more than a yard in diameter, which show the smallest details with a truly admirable clearness.

In the first second of time, your eye receives as much light from the star as does the photographic plate in the same time. But during the course of one hour, the plate receives and stores up about thirty-six hundred times as much light as it or you received in the first instant. There is the secret of the matter. The photographic plate does not only receive, it can also keep and treasure up the light, and that our eyes are not able to do.

If you magnify the amount received in one hour by five or ten, and remember that it is all retained, then you will begin to understand how feeble stars, unseen by man, should become known to us through photography.

CHAPTER XXXII.

THE DAWN OF ASTRONOMY.

THE beginnings of astronomy lie in the distance of earliest historical ages. It is no easy matter to say when first, in all probability, this science grew into existence. Astronomy was alive before the British nation was heard of; before Saxons or Franks had sprung into being; before the Roman Empire extended its iron rule; before the Grecian Empire began to flourish; before the Persian Empire gained power; before the yet earlier Assyrian Empire held sway.

Among the ancient Chaldeans were devoted star-gazers—much more devoted than the ordinary run of educated Anglo-Saxons in this present century. They earnestly sought, in the dim light of those ages, to decipher the meaning of heaven's countless lamps. We have better instruments, and more practiced modes of reasoning; also, we have the collected piles of knowledge built up by our forefathers through tens of centuries. Yet our search is in essence the same as was theirs, to know the truth about the stars: not merely to start some attractive theory, and then to prove that our theory must be right, because we have been so clever as to start it; but to discover that which is in those regions of space. No lower aim than this is worthy to be called scientific.

Mistakes, of course, are made; how should it be otherwise? A man finding his way at night, for the first time, through a wild and unknown country, will almost certainly take some wrong turns before he discovers the right road. In astronomy, as in all other natural sciences, blunders are a necessity if advance is to be made. We have to grope our way to knowledge through observation and conjecture; in simpler terms, through gazing and guessing. Observation of facts leads to conjecture as to the possible causes for those facts; and each conjecture is proved by later observation to be either right or wrong. If proved to be right, it takes its place among accepted truths; if proved to be wrong, it is flung aside. Some pet delusions die hard, because of people's love for them.

Astronomy, as an infant science, mixed up with astrology, existed in the days when the Pyramids were juvenile, and when the Assyrian sculptures were modern. How much farther back who shall say? By astrology, those who studied the heavens endeavored to determine the fate of men and of nations, to predict events, and to interpret results. They paid particular attention to the aspect of the planets, and the general appearance of the firmament.

These observations were at first simple remarks made by the shepherds of the Himalayas after sunset and before sunrise: The phases of the moon, and diurnal retrograde motion of that body with reference to the sun and stars; the apparent motion of the starry sky accomplished silently above our heads; the movements of the beautiful planets through the constellations; the shooting star, which seems to fall from the heavens; eclipses of the sun and moon, mysterious subjects of terror; curious comets, which appear with disheveled hair in the heights of heaven,—such were the first subjects of these old observations made during thousands of years.

The ancient Chaldean star-gazers had rivals among the early Egyptians; and the Chinese profess to have kept actual record of eclipses during between three and four thousand years. The earlier records are not trustworthy; but of thirty-six eclipses reckoned by the Chinese sage, Confucius, no less than thirty-one have been proved true by modern astronomers.

We can not name with certainty those people whose shepherds first gazed with intelligent eyes upon the midnight sky, and noted the steady sweep of stars across the firmament—intelligent eyes, that is to say, so far as man then knew how to use his eyes intelligently; so far as man had begun to note anything in nature, to watch, to compare, and to conjecture causes.

In those times, and indeed for long afterwards, men thought much more about the "influences" of stars upon their own lives, and about supposed prophecies of the future to be read in the sky, than about the actual physical condition of the stars themselves, or the causes and meanings of the various phenomena observed in the heavens.

The desire to know, for the pure delight of knowing, had perhaps hardly begun to dawn upon the mind of man. What people did want to know was what might be going to happen to themselves; whether *they* would be happy or unhappy; and the stars were chiefly of interest as appearing to tell beforehand of troubles or joys to come.

The wisest of men in those times knew less of the

outspread heavens than many a child now knows in our public schools. Earth and sky were one vast bewildering puzzle. They had to discover everything for themselves—how the sun rose each morning and set each evening; how the seasons changed in steady sequence through the year; how the moon and stars journeyed in the night; how the ocean-tides went and came; how numberless every-day phenomena took place.

In olden days the daily rising and setting of the sun was a mystery, accounted for by divers theories, none of which were right; and the march of stars across the midnight sky was a complete puzzle; and an eclipse of sun or moon was a fearful perplexity; and the tides of ocean were a great bewilderment. These things are mysteries still to barbarbous nations; but they perplex us no longer, because we have found out the mode in which such movements, or appearances of movement, are brought about through the action of quite natural causes.

As a first step, in earliest times, the journey of the sun by day, the journeys of moon and stars by night, across the sky, could hardly fail to arrest attention. Very early, too, the stars were grouped into constellations, definite figures and names being attached to each. Many of the constellations are now known to us by names which belong to the earliest historic ages.

The stars were known as "fixed," because they continued unchangeably in their relative positions—that is, in the position held by each star with respect to its neighbor stars—although the whole array of them moved nightly in company; constellations rising and setting at night, as the sun rose and set in the day.

Ages may well have passed before the planets were recognized as distinct from the fixed stars; ages more before any definite plan was noted in their wanderings. In the course of time men's attention was directed to these matters; and one fact after another, of daily or monthly or yearly occurrence, was observed, and commented upon, and became familiar to the minds of people. Very slowly the first beginnings of systematized knowledge took shape and grew, one discovery being made after another, one explanation being offered after another, one theory being started after another.

But an essential difference existed between the infantine science of those primitive days and the matured astronomy of these later days. The whole ancient science was built upon a huge mistake. Men held, as a fact of absolutely unquestionable certainty, that this earth of ours—this small whirling globe, less than eight thousand miles in diameter—was the center, around which sun, moon, and stars all revolved.

The Greek philosopher Thales, who lived about six hundred years before Christ, is said to have laid the foundations of the Grecian astronomy; and Pythagoras is stated to have been one of his disciples. Though, in many respects, Thales wandered wide of the truth, he yet taught many correct ideas; as, for instance, that the stars were made of fire; that the moon had all her light from the sun; that the earth was a sphere in shape, besides other facts respecting the earth's zones and the sun's apparent path in the sky.

He was also one of the three ancient astronomers who were able to calculate and foretell eclipses.

After him came numerous astronomers, of greater or less merit, in the Grecian and in other schools. They watched carefully; they discovered many things of interest; they held divers theories. But one truth never took firm root among them, although several of them dimly apprehended it; and this was the very foundation-truth, for lack of which they were all going hopelessly astray—the simple truth that our earth is *not* the center of the universe, and that our earth *does* move. Yet it is not surprising. No wonder they were slow to grasp such a possibility.

Anything more bewildering to the mind of ancient man than the thought of a solid, substantial world floating in empty space, supported upon nothing, upheld by nothing, can hardly be imagined. As yet little was known of the controlling laws or forces of nature, and that little was with reference only to our earth. The very suspicion of gravitation as a universal law lay in the far-distant future, waiting for the intellect of a Newton to call it out of apparent chaos; and the delicate balance of forces, by means of which the Solar System may almost be said to exist, could not be so much as guessed at.

So men still clung to the thought of earth as the center of all things, and still believed in a little sun, busily circling round her once in every twenty-four hours.

Perhaps the greatest of all ancient astronomers was Hipparchus, about 150 B. C., who did more than any other in those early times to gather together the scattered facts of astronomy, and to arrange them into one united and orderly whole. He it was who discovered the precession of the equinoxes. He studied eclipses and the motions of the various planets. He made elaborate and valuable astronomical tables and star-catalogues; but he, like others, failed to discover the gigantic error which lay at the root of the whole ancient science.

Despite this great mistake, still persistently believed in, and despite the crude notions of early astronomers on many points, it is marvelous how much they did manage to observe and to learn for themselves,—as to the sun and his apparent path, as to the moon and her path, as to the five then known planets and their paths, as to eclipses and other phenomena.

By all such careful watching, although they to some extent missed their aim and fell into mistakes, yet they paved a way to later discoveries and to fuller knowledge. Their work was not thrown away, their trouble was not lost; for out of their very errors grew the fair form of truth.

Nearly three hundred years after the time of Hipparchus came the famous Ptolemy—famous, not, like certain other astronomers, for stupendous genius, or for the brilliancy and accuracy of his observations, but rather noted for the ingenuity of his explanations, and for the adroit manner in which he systematized such knowledge, on the subject of the heavenly bodies, as was then in the possession of mankind.

Ptolemy's name is best known in connection with what is commonly called "The Ptolemaic System of the Universe," and his greatest astronomical work is best known as the *Almagest*.

A great many of Ptolemy's leading notions, as well as the principal mass of facts upon which he worked, were doubtless borrowed from Hipparchus. The latter is said to have explained the movements of the sun and of the moon by means of small epicycles, traveling round the earth on circular orbits; and even Hipparchus did not originate this fundamental idea of the so-called "Ptolemaic System," which had indeed been held by the ancients in much earlier days. Hipparchus worked it out to some extent, and Ptolemy carried on the process much farther, giving forth the system to the world in such wise that ever since it has gone by his name.

The theory of cycles and epicycles lasted long—lasted from the time of Ptolemy in the second century to the time of Copernicus in the sixteenth century, of whom we have already spoken.

[For about two thousand years, astronomers observed attentively the apparent revolutions of the heavenly bodies, and this attentive study gradually showed them a large number of irregularities and inexplicable complications, until at last they recognized that they were deceived as to the earth's position, in the same way that they had been deceived as to its stability. The immortal Copernicus, in particular, discussed with perseverance the earth's motion, already previously suspected for two thousand years, but always rejected by man's self-love; and when this learned Polish canon bid adieu to our world in the year 1543, he bequeathed to science his great work, which demonstrated clearly the long-standing error of mankind.—*F.*]

The new theory of Copernicus had to make its way slowly against the dead weight of unreasoning public opinion, and against wrongly-reasoning hierarchical opposition. In time, gradually, despite all resistance, the truth made its way, and was generally received, though not till Galileo had been forced by ecclesiastical authority to recant his opinions. All the same, Copernicus, Kepler, Galileo, were right; public sentiment and the Church were wrong,—the earth did move, and was not the fixed universe center, and, by and by, those who lived on earth had to acknowledge the same.

The succession of these great men is interesting to note. Between the two shining lights, Hipparchus and Ptolemy, nearly three hundred years intervened. But as the history of the world advances, we find brilliant scientific minds appearing more quickly, one following close upon another, instead of their being divided by long intervals of blankness.

Within thirty years from the death of Copernicus, were born two mighty men of Science: Kepler, who lived till 1630; and Galileo, who lived till 1642, when England was plunged in civil strife.

Between Copernicus and Kepler came Tycho Brahe. Copernicus was a Roman ecclesiastic, born in Poland. Tycho Brahe was a wealthy Danish nobleman, an ardent lover of astronomy, and a most patient and accurate observer. But Tycho held fast by the old astronomy. He was not convinced by the arguments of Copernicus. To him, earth was still the motionless center of all things. He was willing to allow that the rest of the planets circled round the sun, as an explanation of things which he could not help seeing; but the sun itself had still, in Tycho's imagination, to revolve daily, with planets and stars and the whole sky, round our earth.

As an explainer of causes, therefore, Tycho rose to no lofty heights. The real good which he did was in watching and noting actual phenomena, not in trying to explain how those phenomena were brought about. This was left for one of his young pupils, a sickly German lad, named John Kepler.

In later years, Kepler made a grand use of his master's mass of careful and thoroughly dependable observations. These had really been gathered together by Tycho, with a view of disproving the Copernican theory, and of establishing the main features of the olden astronomy, with certain improvements. But Kepler used the accumulated information to disprove the old astronomy, and to establish the truth of the new Copernican system which Tycho had rejected. He found a vast advantage, ready to his hand, in the collection of careful observations systematically worked out by Tycho in his observatory during many a long year. And Kepler did not fail to use his advantage.

The planets still refused, as they always had refused, to keep to the paths marked out for them by astronomers. Kepler grappled with the difficulty. He particularly turned his attention to Mars, that near neighbor of ours, in which we are all now so much interested. A long series of close observations of Mars's movements, made by Tycho in the course of many years, lay before him; and he knew that he could depend upon the absolute honesty and accuracy of Tycho's work.

With immense labor and immense patience, he went into the matter, still clinging to the old idea of a circular pathway round the sun. He tried things this way and that way. He placed his orbit in imagination after one mode and another mode; he conjectured various arrangements; he tested and proved them in turn, comparing each plan with observations made,—and still Mars defied all his efforts; still the little world went persistently wrong, traveling contrary to every theory.

Thus far, nobody had ever thought of an orbit of any other shape than a circle. When a straightforward journey round a circle was found impossible, then epicycles were introduced to explain the planet's perplexing motions. No one had dreamt of an elliptic pathway. But suddenly a gleam of daylight came upon this groping in the dark. What if Mars traveled round the sun—*not* in a circle, but in an oval?

Kepler tried this oval and that oval, comparing observations past—testing, examining, proving or disproving, with unwearied patience. And at length he found, beyond dispute, that Mars actually did travel round the sun in a yearly pathway, which was shaped, not as a circle, but as the kind of oval known as an *ellipse;* and, further, that the sun was not in the exact center of this ellipse, but to one side of the center, in one of the foci. Kepler's success in detecting the true shape of a planetary orbit is doubtless owing to the fact that the orbit of Mars makes a wider departure from the circular form than any of the other important planets.

One step further would have led Kepler to the knowledge that comets also travel in elliptic orbits—only in very long and narrow ones generally. But he had quite made up his mind that all comets merely paid our sun one visit, and never by any chance returned; so he did not trouble himself to enter into calculations with respect to them. This taking for granted of ideas long held was a very common practice in those days.

The above wonderful discovery of elliptic orbits, in place of circular orbits, was only one among many made by the illustrious Kepler—discoveries not carelessly hit upon, as one might pick up in the street a valuable stone which somebody had dropped, but earnestly sought for, and found with toil and diligence, as gems are first found in foreign lands, after much seeking and long patience. Casual discoveries in science are rare. Attainment is far more usually made through intense study, through hard work, through profound thought, through a gradual groping upward out of darkness to the point where daylight breaks forth.

The three well-known "Laws of Kepler" still lie at the very foundation of the modern system of astronomy. They are: First, every planet moves in an elliptical orbit, in one focus of which the sun is situated; second, the line drawn from the sun to a planet moves over equal areas in equal times; and, third, the square of the periodic times of the planets varies inversely as the cubes of their distances.

CHAPTER XXXIII.

MODERN ASTRONOMY.

STILL, however, the world was not convinced of the truth of the Copernican system. One man here or there saw and acknowledged its reality. The mass of mankind continued firmly to believe that they dwelt at the universe center, and utterly to scout the idea of a moving earth. Indeed, it may be very much doubted whether, not merely the mass of mankind, but even the most civilized and cultivated portion of that mass, had as yet, to any wide extent, heard a whisper of the new theories. Knowledge, in those days, traveled from place to place with exceeding deliberation.

Another great man arose, contemporary with Kepler—a man whose story seizes more strongly upon our imagination than that of the severe and successful studies of Kepler. From an ordinary worldly point of view, Kepler would hardly be looked upon as altogether a successful man. He was poor, and in sickly health. His chief book was promptly prohibited by Rome—then a power in the whole reading world of Europe to an extent which can hardly be realized in these days of freedom. Kepler's book was placed in the forbidden category, side by side with the dying publication of Copernicus. This checked all hope of literary gains, and life with Kepler was one long struggle. But if he could have looked ahead two or three hundred years; if he could have foreseen the time when not a few wise men of science alone, but all the civilized world, including the very Church which condemned him, should meekly have accepted and indorsed his discoveries,—then he would have been able to gauge the measure of his own real success.

Moreover, with all his troubles, Kepler was so happy as to escape absolute persecution. He lived in a country which knew something of liberty as an ideal; he was left to write and work freely; and, at least, no public recantation of what he held to be truth was forced from him.

Galileo's was a sadder tale. A Florentine of noble birth, he turned his attention early to science as the needle turns to the north, and before the age of twenty-seven he already occupied a foremost position as mathematical lecturer at Pisa. There it was that he direfully offended the philosophers of the day by proving Aristotle to be in the wrong. For Aristotle was the great master of the schoolmen. His philosophy and theories were accepted as final, and whatsoever could not be proved in accordance with them was regarded as untrue. Aristotle had declared that if two weights—the one being ten times as heavy as the other—were dropped together from a height, the heavier weight would reach the ground ten times as quickly as the lighter.

Everybody believed this, and nobody had ever thought of putting it to the test of actual trial. Aristotle had said so, and Aristotle was indorsed as a whole by all the schools of Europe; and what more could possibly be wanted by any reasonable human being?

PROMINENT ASTRONOMERS OF THE NINETEENTH CENTURY.

But Galileo was not so easily satisfied. He looked into the matter independently, bent upon finding out, not what Aristotle had thought or what the Church might support, but simply what actually *was*.

Then, in view of many observers, called together for the occasion, he dropped, at the same instant, from the top of the leaning tower of Pisa, a shot weighing one pound and another shot weighing a hundred pounds. And, behold! both reached the ground simultaneously. There was not a fraction of difference between the two. To prove Aristotle wrong was to prove everybody wrong, the authorities of the Church included. Men calmly declined to believe their own eyes, and Galileo was in very evil odor.

From falling bodies he went on to subjects of yet wider interest to people in general. He read about and eagerly embraced the Copernican system. Not only did he accept it himself, but he vigorously set to work to teach and convince others also. This, as might only be expected, aroused vehement opposition. But Galileo was still in the heyday of his powers, and he was not to be easily silenced. In those days he could afford to press onward in the teeth of resistance, and to laugh at men who would not listen.

When at Venice, a report reached him of a certain optician in Holland who had happened to hit upon a curious "combination of lenses," through which objects afar off could be actually seen as if they were near. Galileo promptly seized upon the notion, and within twenty-four hours he had rigged up a small, rough telescope out of an old organ-pipe and two spectacle-glasses.

Thus the first telescope was made—a very elementary affair, formed of two lenses, convex and concave, and capable of magnifying some three times—a mere child's toy compared with telescopes of the present day, but of exceeding value and interest, because it was the very first; because in all the history of the world no telescope had ever been manufactured before.

This earliest effort was speedily improved upon. The next trial produced a tube which could magnify seven or eight times; and in a little while, Galileo had a telescope magnifying as much as thirty-two times. Even thirty-two times does not sound very startling to us; but in those days the revelations of such an instrument came upon men like a thunderclap.

Through it could be seen clearly the broken nature of the moon's surface—the craters, the shadows, the mountains. Through it could be seen the phases of Venus, which no man had ever yet detected, but which Copernicus had declared must certainly exist, if indeed his system were a reality. Through it could be seen four of the moons of Jupiter, and their ceaseless journeyings. Through it could be seen spots upon the face of the sun, the movements of which showed the rotation of the sun upon his axis. Through it could be seen the very curious shape of Saturn, caused by his rings, though a much stronger power was needed to separate the rings from the body of the planet.

Of all these discoveries, and many others also made by Galileo, none seemed worse to the bigoted schoolmen of his day than the preposterous notion of black spots upon the sun's face.

Rather curiously, no human being seems ever before then to have noticed such spots, though often they are quite large enough to be detected by the naked eye. A general belief had been held that the sun was and must be perfect—an absolutely spotless and unblemished being, the purest symbol of celestial incorruptibility—and the said discovery went right in the teeth of deductions drawn from, and lessons founded upon, this belief. Therefore the schoolmen held out and determinedly resisted, and would have nought to do with Galileo or his telescopes, shutting their eyes to marvels newly revealed.

But Kepler was a true scientist, a man who earnestly pursued truth, and sought to discover it at any hazard. He and Galileo were contemporaries, and they held communication on these subjects. Some of Galileo's discoveries went quite as much in the teeth of some of Kepler's most dearly-loved theories as they did in the teeth of the schoolmen's most dearly-loved doctrines. Yet, none the less, Kepler welcomed them with great delight and eagerness, showing thereby his true greatness of character.

Till after the age of fifty, Galileo was allowed to go on undisturbed, or at least not materially disturbed, by opposition in his career of success—observing, learning, theorizing, calculating, explaining, teaching—a prominent figure indeed. Not only did he take a large share in establishing firmly the Copernican system of astronomy, but some of the principal laws of motion were discovered by him. He first found the uses of a pendulum, and there is strong reason for believing that the first microscope, as well as the first telescope, was from his hands.

But after the age of fifty, reverses came, and one blow succeeded another. Rome had been gradually waking up to the true sense of all that was implied by his discoveries and his teaching, and at length she let loose her thunders—impotent thunders enough, so far as regarded any permanent checking of the progress of truth, but by no means impotent to crush one defenseless victim, the champion of scientific truth in that country and age.

Years of greater or less opposition, amounting often to persecution, were followed by an imperious order commanding him to Rome. There he was examined and severely "questioned," and recantation was forced from him, worth as much as such forced recantations commonly are worth.

Even then he did not cease from the work that he loved; even in prison he carried it on; even when he might not write, nothing could stop him from thinking. But his health was broken; his liberty was taken away; illness followed illness; his favorite daughter died; his eyesight gradually failed; the last book that he wrote might not for a long while be printed; and at last he passed away, mercifully set free from those who, in an essentially bigoted and intolerant age, were incapable of appreciating his greatness.

They would fain *then* have denied him even a respectable grave. Now, all the civilized world unites to honor the name of Galileo.

CHAPTER XXXIII.

ISAAC NEWTON.

EVEN the great fame of Galileo must be relegated to a second place in comparison with that of the philosopher who first saw the light in a Lincolnshire farmhouse on the 25th of December, 1642, the very same year of Galileo's death. His father, Isaac Newton, had died before his birth. The little Isaac was at first so frail and weakly that his life was despaired of. The watchful mother, however, tended her delicate child with such success that he ultimately acquired a frame strong enough to outlast the ordinary span of human life.

In due time the boy was sent to the public school at Grantham. That he might be near his work, he was boarded at the house of Mr. Clark, an apothecary. He had at first a very low place in the class. His first incentive to diligent study seems to have been derived from the circumstance that he was severely kicked by one of the boys above him in the class. This indignity had the effect of stimulating young Newton's activity to such an extent that he not only attained the desired object of passing over the head of the boy who had maltreated him, but continued to rise until he became the head of the school.

The play-hours of the great philosopher were devoted to pursuits very different from those of most schoolboys. His chief amusement was found in making mechanical toys, and various ingenious contrivances. He watched, day by day, with great interest, the workmen engaged in constructing a windmill in the neighborhood of the school, the result of which was that the boy made a working model of the windmill and of its machinery, which seems to have been much admired as indicating his aptitude for mechanics. We are told that he also indulged in somewhat higher flights of mechanical enterprise. The first philosophical instrument he made was a clock, which was actuated by water. He also devoted much attention to the construction of paper kites, and his skill in this respect was highly appreciated by his schoolfellows. Like a true philosopher, even at this stage, he experimented on the best methods of attaching the string, and on the proportions which the tail ought to have.

When the schoolboy at Grantham was fourteen years of age, it was thought necessary to recall him from the school. His recently-born industry had been such that he had already made good progress in his studies, and his mother hoped that he would now lay aside his books, and those silent meditations to which, even at this early age, he had become addicted. It was hoped that instead of such pursuits, which were deemed quite useless, the boy would enter busily into the duties of the farm and the details of a country life. But before long it became manifest that the study of nature and the pursuit of knowledge had such a fascination for the youth that he could give little attention to aught else. It was plain that he would make but an indifferent farmer. He greatly preferred experimenting on water-wheels, and working at mathematics in the shadow of a hedge.

Fortunately for humanity, his mother, like a wise woman, determined to let her boy's genius have the scope which it required. He was accordingly sent back to Grantham school, with the object of being trained in the knowledge which would fit him for entering the University of Cambridge.

It was the 5th of June, 1660, when Isaac Newton, a youth of eighteen, was enrolled as an undergraduate of Trinity College, Cambridge. Little did those who sent him there dream that this boy was destined to be the most illustrious student who ever entered the portals of that great seat of learning. Little could the youth himself have foreseen that the rooms near the gateway which he occupied would acquire a celebrity from the fact that he dwelt in them, or that the antechapel of his college was in good time to be adorned by that noble statue of himself which is regarded as the chief art treasure of Cambridge University, both on account of its intrinsic beauty and the fact that it commemorates the fame of her most distinguished alumnus, Isaac Newton, the immortal astronomer. His advent at the university seemed to have been by no means auspicious or brilliant. His birth was, as we have seen, comparatively obscure, and though he had already given indication of his capacity for reflecting on philosophical matters, yet he seems to have been but ill-equipped with the routine knowledge which youths are generally expected to take with them to the universities.

From the outset of his college career, Newton's attention seems to have been mainly directed to mathematics. Here he began to give evidence of that marvelous insight into the deep secrets of nature which, more than a century later, led so dispassionate a judge as Laplace to pronounce Newton's immortal work as pre-eminent above all the productions of the human intellect. But though Newton was one of the very greatest mathematicians that ever lived, he was never a mathematician for the mere sake of mathematics. He employed his mathematics as an instrument for discovering the laws of nature.

His industry and genius soon brought him under the notice of the university authorities. It is stated in the university records that he obtained a scholarship in 1664. Two years later we find that Newton, as well as many residents in the university, had to leave Cambridge temporarily on account of the breaking out of the plague. The philosopher retired for a season to his old home at Woolsthorpe, and there he remained until he was appointed a Fellow of Trinity College, Cambridge, in 1667. From this time onwards, Newton's reputation as a mathematician and as a natural philosopher steadily advanced, so that in 1669, while still but twenty-seven years of age, he was appointed to the distinguished position of professor of Mathematics at Cambridge. Here he found the opportunity to continue and develop that marvelous career of discovery which formed his life's work.

The earliest of Newton's great achievements in natural philosophy was his detection of the composite character of light. That a beam of ordinary sunlight is, in fact, a mixture of a very great number of different colored lights, is a doctrine now familiar to every one who has the slightest education in physical science. We must, however, remember that this discovery was really a tremendous advance in knowledge at the time when Newton announced it.

A certain story is often told about Newton. It is

SIR ISAAC NEWTON.

said that as he sat in a garden an apple fell from a tree, and that its fall set him thinking, and that, in consequence of this train of thought, he found out all about the law of attraction or gravitation.

But the fall of an apple could not possibly have set Newton thinking, because he had already been thinking long and hard upon these difficult questions. Indeed, it was *because* he had been so thinking, *because* his mind was in a watchful and receptive condition, that so slight a matter as a falling apple should, perhaps, have suggested to him a useful train of ideas.

Newton was a man who really did think, and who really did think intensely. He was not, like the ordinary run of people, one who merely had a good many notions walking loosely through his brain. He would bend his mind to a subject and be absolutely absorbed in it; later in life so absorbed that he would forget to finish his dressing, and would be unable to say whether or no he had dined. This sort of forgetfulness does sometimes spring from vacancy of mind; but with Newton it arose from fullness of thought.

An apple falls, of course, as a stone or any other body falls, because the earth attracts it. Otherwise, it might as easily rise and float away into the sky. But this was known perfectly well long before Newton's days. It was clearly understood that the earth had a curious power of drawing all things towards herself. People could not explain why or how it was so; neither can we explain now; but they were fully aware of the fact.

So men knew something of the force called gravity; but they knew it mainly, almost exclusively, as having to do with our earth, and with the things upon our earth. They had not begun definitely to think of gravity as having to do with bodies in the heavens; as being a chief controlling force in the whole Solar System; much less as reigning throughout the vast universe of stars. It had not come into their minds with any distinctness that, just as the earth pulls towards her own center a mountain, a horse, a man, so the sun pulls toward his center the earth, the moon, and all the planets.

Certain glimmerings of the notion had, indeed, begun to dawn on the horizon of learning. Horrocks, Borelli, Robert Hook, Wren, Halley, and others had *glimpses* of univeral gravitation; but only glimpses. A gifted English philospher, late in the sixteenth century, Gilbert by name, went so far as to speak of earth and moon acting one upon the other "like two magnets." He also saw the effect of the moon's attraction in bringing about ocean tides, though he oddly explained that the said attraction was not so much exerted upon ocean waters as upon "subterranean spirits and humors"—whatever he may have meant by such an expression.

Kepler, not long after Gilbert, made further advance. He gained some definite notions as to the nature of gravity; and he saw distinctly that the moon's attraction was the main cause of the tides. The following remarkable statement is found in his writings: "If the earth ceased to attract its waters, the whole sea would mount up and unite itself with the moon. The sphere of the attracting force of the moon extends even to the earth, and draws the waters towards the torrid zone, so that they rise to the point which has the moon in the zenith." Here is a distinct enough grasp of the fact that gravity can and does act through distance, between worlds separated by thousands of miles.

Yet, though both Gilbert and Kepler saw so much,

they failed to go farther. Neither of the two had any clear knowledge of the modes in which gravitation acts, or of the laws which govern its working. Gilbert and Kepler alike, while dimly recognizing, not only the moon's attraction for earth's oceans, but the mutual attraction of earth and moon each for the other, were utterly perplexed as to what force could possibly hold the two bodies asunder. Gilbert could seriously talk of "subterranean spirits and humors." Kepler could soberly write of an "animal force" or some tantamount power in the moon, which prevented her nearer approach to earth.

Galileo was equally vague on the subject. With all his brilliant gifts, and while acknowledging the earth's attractive power over the moon, he would not even admit the existence of *mutual* attraction, and talked of the moon as being compelled to "follow the earth." Unquestionably there are times when the moon does follow the earth, and that is when the earth happens to go before. But quite as often it happens that the moon goes before, and the earth follows after.

All these great men were groping near the truth, yet none managed to find the clue that was wanted. They left the unfinished process of thought to be carried on by the master-mind of Newton. And Newton set himself to think the question out. He used Kepler's observations. He sought to penetrate the secrets of a Solar System in mysterious and inviolable order. Some reason or cause for that order he knew must exist—if only it might be grasped! He pondered deeply; lost in profound cogitation, through stormy days of civil strife, of plague and loss and suffering. The object of his life was to discover, if so he might, what power or what forces retained the different worlds in their respective pathways; in short, what manner of celestial guidance was vouchsafed to the planets.

For although men had now learned the shape of the planet-orbits, and even knew certain laws which helped to govern the planet-motions, they had not discovered one supreme controlling law. They did not dream of gravity in the heavens as a prevailing force. There was absolutely no reason, with which men were acquainted, why each planet should preserve its own particular distance from the sun; why Mars should not wander as far away as Jupiter; why Mercury should not flee to the position of Saturn; why our own earth should be always about as near as she always is.

The resources of Newton's genius seemed equal to almost any demand that could be made upon it. He saw that each planet must disturb the other, and in that way he was able to render a satisfactory account of certain phenomena which had perplexed all preceding investigators. That mysterious movement by which the pole of the earth sways about among the stars, had been long an unsolved enigma; but Newton showed that the moon grasped with its attraction the protuberant mass at the equatorial regions of the earth, and thus tilted the earth's axis in a way that accounted for the phenomenon which had been known, but had never been explained, for two thousand years. All these discoveries were brought together in that immortal work, Newton's "Principia," the greatest work on science that the world had yet seen.

It may well be that, as Newton pondered this perplexing question while seated in the garden, an apple dropping to the ground might have sufficed to turn his wide-awake and ready mind in the direction of that familiar force, gravity, which draws all loose bodies towards earth's surface. The apple fell because of gravity. Quite naturally the question might have arisen in Newton's mind—"Why not this same gravity in the heavens also? If the earth draws an apple towards herself, why should not the earth draw the moon? Why should not the sun draw the earth? Nay, more, why should not the sun draw all the planets?" Newton set himself to work out this problem, allowing for the distance of the moon from earth, and for the bulk and weight of both earth and moon.

And because the answer was *not* precisely what he had looked for, he put the calculation aside, and waited for years. It seems that he mentioned his conjecture to no living person. After many years he took it up again, and worked it out afresh. By this time new measurements of earth's surface, and new calculations of earth's size had been made, and Newton had fresh figures to go upon. This time the answer came just as he had originally hoped, and the truth of his surmise was thus proved.

Gravity held the moon near the earth, preventing her from wandering off into unknown distances. Gravity held earth and moon, Mars, Venus, Jupiter, each planet, in its own pathway, within a certain distance of the sun. Gravity held many of the comets as permanent members of the Solar System. So strong, indeed, was the drawing of gravity found to be, that only the resisting force of each bright world's rapid rush round the sun could keep the sun and his planets apart.

Newton had discoved the long-sought secret. He had found that "every particle of matter in the universe attracts every other particle." He had found also the main laws which govern the working of that attraction—that it depends, first, upon the mass or amount of "matter" in each body; that it depends, secondly, upon the distance of one body from another. He found, too, how far attraction increases with greater nearness, and decreases with greater distance. All these and countless other questions he worked out in the course of years. But first he had grasped the great leading principle, after which men had until then groped in vain, that by the force of gravity the sun and his worlds and their moons are all bound together into one large united family or system.

Though Newton lived long enough to receive the honor that his astonishing discoveries so justly merited, and though for many years of his life his renown was much greater than that of any of his contemporaries, yet it is not too much to say that, in the years which have since elapsed, Newton's fame has been ever steadily advancing.

We hardly know whether to admire more the sublime discoveries at which he arrived, or the extraordinary character of the intellectual processes by which those discoveries were reached. He died on Monday, March 20, 1727, in the eighty-fifth year of his age.

CHAPTER XXXV.

LATER ASTRONOMY.

SIR WILLIAM HERSCHEL was born in the city of Hanover, November 15, 1738. He was the second son of Isaac Herschel, a musician, who brought him up, with his four other sons, to his own profession.

A faithful chronicler has given us an interesting account of the way in which Isaac Herschel educated his boys; the narrative is taken from the recollections of one who, at the time, was a little girl of five or six years old. She writes:

"My brothers were often introduced as solo performers and assistants in the orchestra at the court, and I remember that I was frequently prevented from going to sleep by the lively criticisms on music on coming from a concert. Often I would keep myself awake that I might listen to their animated remarks, for it made me happy to see them so happy. But generally their conversation would branch out on philosophical subjects, when my brother William and my father often argued with such warmth that my mother's interference became necessary when the names Euler, Leibnitz, and Newton sounded rather too loud for the repose of her little ones, who had to be at school by seven in the morning."

The child, whose reminisences are here given, became afterwards the famous Caroline Herschel. The narrative of her life is a most interesting book, not only for the account it contains of the remarkable woman herself, but also because it presents the best picture we have of the great astronomer, to whom Caroline devoted her life.

This modest family circle was, in a measure, dispersed at the outbreak of the Seven Years' War in 1756. The French proceeded to invade Hanover, which, it will be remembered, belonged at this time to the British dominions. Young William Herschel had already obtained the position of a regular performer in the regimental band of the Hanoverian Guards, and it was his fortune to obtain some experience of actual warfare in the disastrous battle of Hastenbeck. He was not wounded, but he had to spend the night after the battle in a ditch, and his meditations on the occasion convinced him that soldiering was not the profession exactly suited to his tastes. He left his regiment by the very simple, but somewhat risky, process of desertion. He had, it would seem, to adopt disguises in order to effect his escape. By some means he succeeded in eluding detection, and reached England in safety.

The young musician must have had some difficulty in providing for his maintenance during the first few years of his abode in England.

It was not until he had reached the age of twenty-two that he succeeded in obtaining any regular appointment. He was then made instructor of music to the Durham Militia. Shortly afterwards, his talents being more widely recognized, he was appointed as organist of the parish Church at Halifax.

In 1766 we find that Herschel had received the

further promotion of organist in the Octagon Chapel, at Bath. Bath was then, as now, a highly fashionable resort, and many notable personages patronized the rising musician. Herschel had other points in his favor besides his professional skill; his appearance was striking, his address superb, and his conversation animated, interesting, and instructive; and even his nationality was a distinct advantage, inasmuch as he was a Hanoverian in the reign of King George the Third. From his earliest youth, Herschel had been endowed with that invaluable characteristic, an intense desire for knowledge. He naturally wished to perfect himself in the theory of music, and thus he was led to study mathematics. When he had once tasted the charms of mathematics, he saw vast regions of knowledge unfolded before him, and in this way he was induced to direct his attention to astronomy. More and more this pursuit engrossed his attention until, at last, it had become an absorbing passion.

It was with quite a small telescope, which had been lent him by a friend, that Herschel commenced his career as an observer. However, he speedily discovered that to see all he wanted to see, a telescope of far greater power would be necessary.

He commissioned a friend to procure for him in London a telescope with high magnifying power. Fortunately for science the price was so great that it precluded the purchase, and he set himself at work to construct one. After many trials he succeeded in making a reflecting instrument of five feet focal length, with which he was able to observe the rings of Saturn and the satellites of Jupiter.

It was in 1774, when the astronomer was thirty-six years old, that he obtained his first glimpse of the stars with an instrument of his own construction. Night after night, as soon as his musical labors were ended, his telescopes were brought out.

His sister Caroline, who occupies such a distinct place in scientific history, the same little girl to whom we have already referred, was his able assistant, and when mathematical work had to be done, she was ready for it. She had taught herself sufficiently to enable her to perform the kind of calculations—not, perhaps, very difficult ones—that Herschel's work required; indeed, it is not much to say that the mighty life-work which this man was enabled to perform could never have been accomplished had it not been for the self-sacrifice of this ever-loving and faithful sister.

It was not until 1782 that the great achievement took place by which Herschel at once sprang into fame. He appears to have formed a project for making a close examination of all the stars above a certain magnitude for the purpose of discovering, if possible, a parallax among them. Star after star was brought to the center of the field of view of his telescope, and after being carefully examined, was then displaced, while another star was brought forward to be submitted to the same process.

In the great review which Herschel undertook, he doubtless examined hundreds, or perhaps thousands of stars, allowing them to pass away without note or comment. But on an ever-memorable night in March, 1782, it happened that he was pursuing his task among the stars in the constellation of Gemini. One star was noticed which, to Herschel's acute vision, seemed different from the stars which in so many thousands are strewn over the sky. There was something in the starlike object that immediately arrested his attention, and made him apply to it a higher magnifying power. This at once disclosed the fact that the object possessed a disk—that is, a definite, measurable size—and that it was thus totally different from any of the hundreds and thousands of stars which exist elsewhere in space. The organist at the Octagon Chapel at Bath had discovered a new planet with his home-made telescope. Great was the astonishment of the scientific world when the Bath organist announced that the five planets, which had been known from all antiquity, must now admit the company of a sixth.

The now great astronomer was invited to Windsor, and to bring his famous telescope in order to exhibit the planet to George III, and to tell his majesty all about it. The king took so great a fancy to Herschel that he proposed to confer on him the title of "his majesty's own astronomer," to assign him a residence near Windsor, to provide him with a salary, and to furnish such funds as might be required for the erection of great telescopes, and for the conduct of that mighty scheme of celestial observation on which Herschel was so eager to enter.

No single discovery of Herschel's later years was, however, of the same momentous description as that which first brought him to fame. There is no separate collection of his writings, and very scanty accounts of his life have been published; but he who has written his name among the stars needs no other testimonial to his fame. Prior to his time the number of bodies known as belonging to the Solar System was eighteen, including secondary planets and Halley's comet. To these he added nine; namely, Uranus and six satellites, and two satellites of Saturn.

Though no additional honors could add to his fame, Dr. Herschel, in 1816, received the decoration of the Guelphic Order of Knighthood. In 1820 he was elected president of the Astronomical Society, and among their Transactions, the next year, he published an interesting memoir on the places of one hundred and forty-five double-stars. This paper was the last which he lived to publish. His health had begun to decline, and on the 24th of August, 1822, he sank under the infirmities of age, having completed his eighty-fourth year. He was survived by his widow, Lady Herschel, an only son, Sir John F. W. Herschel, and his sister Caroline, who died in 1847, in her ninety-eighth year.

LAPLACE.

The author of "Celestial Mechanics" was born at Beaumont-en-Auge in 1749, just thirteen years later than his renowned friend Lagrange. His father was a farmer, but appears to have been in a position to provide a good education for a son who seemed promising. The subject which first claimed his attention was theology. He was, however, soon introduced to the study of mathematics, in which he presently became so proficient that, while he was still no more than eighteen years old, he obtained employment as a mathematical teacher in his native town.

Desiring wider opportunities for study, young Laplace started for Paris, being provided with letters of introduction to D'Alembert, who then occupied the most prominent positions as a mathematician in France, if not in Europe. On presenting these letters, he seems to have had no reply; whereupon, Laplace wrote to D'Alembert, submitting a discussion of some point in dynamics.

This letter instantly produced the desired effect. D'Alembert thought that such mathematical talent as the young man displayed was in itself the best of introductions to his favor. It could not be overlooked, and accordingly he invited Laplace to come and see him. Laplace, of course, presented himself, and erelong D'Alembert obtained for the rising philosopher a professorship of Mathematics in the military school in Paris. This gave the brilliant young mathematician the opening for which he sought, and he quickly availed himself of it.

Laplace's most famous work is "Celestial Mechanics," in which he essayed a comprehensive attempt to carry out, in much greater detail, the principles which Newton had laid down. The fact was that Newton had not only to construct the theory of gravitation, but he had to invent the mathematical processes by which his theory could be applied to the explanation of the movements of the heavenly bodies. In the course of the century which had elapsed between the time of Newton and the time of Laplace, mathematics had been extensively developed. The disturbances which one planet exercises upon the rest can only be fully ascertained by the aid of long calculations, and for these calculations analytical methods are required.

With an armament of mathematical methods which had been perfected since the days of Newton by the labors of two or three generations of mathematical inventors, Laplace essayed in his "Celestial Mechanics" to unravel the mysteries of the heavens. It will hardly be disputed that the book which he has produced is one of the most difficult books to understand that has ever been written. The investigations of Laplace are, generally speaking, of too technical a character to make it possible to set forth any account of them in such a work as the present. He did, however, publish one treatise, called the "System of the Universe," in which, without introducing mathematical symbols, he was able to give a general account of the theories of the celestial movements, and of the discoveries to which he and others had been led. In this work Laplace laid down the principles of the nebular theory, which, in modern days, has been generally accepted.

The nebular theory gives a physical account of the origin of the Solar System, and has already been explained in this volume. Each advance in science seems to make it more certain that this hypothesis substantially represents the way in which our Solar System has grown to its present form.

Not satisfied with a career which was merely scientific, Laplace sought to connect himself with public affairs. Napoleon appreciated his genius, and desired to enlist him in the service of the state. Laplace was appointed to the office of minister of the interior. The experiment was not successful, for he was not by nature a statesman. In despair of Laplace's capacity as an administrator, Napoleon declared that he carried the spirit of his infinitesimal calculus into the management of business. Indeed, Laplace's political conduct hardly admits of much defense. While he accepted the honors which Napoleon showered on him in the time of his prosperity, he seems to have forgotten all this when Napoleon could no longer render him service. Laplace was made a marquis by Louis XVIII. During the latter part of his life the philosopher lived in a retired country-place at Arcueile. Here he pursued his studies, and, by strict abstemiousness, preserved himself from many of the infirmities of old age.

He was endowed with remarkable scientific sagacity; but above all his powers his wonderful memory shone pre-eminent. His "Celestial Mechanics" is, next to Newton's "Principia," the greatest of astronomical works.

He died on March 5, 1827, in his seventy-eighth year, his last words being, "What we know is but little, what we do not know is immense."

LEVERRIER.

We are apt to identify the idea of an astronomer with that of a man who looks through a telescope at the stars; but the word astronomer has really a much wider significance. No man who ever lived has been more entitled to be designated an astronomer than Urbain Jean Joseph Leverrier, and yet it is certain that he never made a telescopic discovery of any kind. In mathematics, however, he excelled, and he simply used the observations of others as the basis of his calculations.

These observations form, as it were, the raw material on which the mathematician exercises his skill. It is for him to elicit from the observed places the true laws which govern the movements of the heavenly bodies. Here is indeed a task in which the highest powers of the human intellect may be worthily employed. To Leverrier it has been given to provide a superb illustration of the success with which the mind of man can penetrate the deep things of nature.

The illustrious Frenchman was born on the 11th of March, 1811, at Saint-Lô, in the department of Manche. In the famous polytechnic school for education in the higher branches of science he acquired considerable fame as a mathematician. His labors at school had revealed to Leverrier that he was endowed with the powers requisite for dealing with the subtlest instruments of mathematical analysis. When he was twenty-eight years old his first great astronomical investigation was brought forth. This was the profound calculation of the disturbances of the planet Uranus and the causes of them.

The talent which his researches displayed brought Leverrier into notice. At that time the Paris Observatory was presided over by Arago, a savant who occupies a distinguished position in French scientific annals. Arago at once perceived that Leverrier possessed the qualifications suitable for undertaking a problem of great importance and difficulty that had begun to force itself on the attention of astronomers. What this great problem was, and how astonishing was the solution it received, must now be considered.

Ever since Herschel brought himself into fame by the discovery of Uranus, the movements of this new addition to the Solar System had been scrutinized with care and attention. The position of Uranus was thus accurately determined from time to time. When due allowance was made for whatever influence the attraction of Jupiter and Saturn and all the other planets could possibly produce, the movements of Uranus were still inexplicable. It was perfectly obvious that there must be some other influence at work besides that which could be attributed to the planets already known.

Astronomers could only recognize one solution of such a difficulty. It was impossible to doubt that there must be some other planet in addition to the bodies at that time known, and that the perturbations of Uranus, hitherto unaccounted for, were due to the disturbances caused by the action of this unknown planet. Arago urged Leverrier to undertake the great problem of searching for this body. But the conditions of the search were such that it must be conducted on principles wholly different from any search which had ever before been undertaken for a celestial object. For this was not a case in which mere survey with a telescope might be expected to lead to the discovery.

There are in the heavens many millions of stars, and the problem of identifying the planet, if indeed it should lie among these stars, seemed a very complex matter. Of course, it is the abundant presence of the stars which causes the difficulty.

The materials available to the mathematician for

the solution of this problem were to be derived solely from the discrepancies between the calculated places in which Uranus should be found, taking into account the known causes of disturbances, and the actual places in which observation had shown the planet to exist. Here was, indeed, an unprecedented problem, and one of extraordinary difficulty. Leverrier, however, faced it, and, to the astonishment of the world, succeeded in carrying it through to a brilliant solution.

After many trials, Leverrier ascertained that, by assuming a certain size, shape, and position for the unknown planet's orbit, and a certain value for the mass of the hypothetical body, it would be possible to account for the observed disturbances of Uranus. Gradually it became clear to his perception, not only that the difficulties in the movements of Uranus could be thus explained, but that no other explanation need be sought for. And now for an episode in this history which will be celebrated so long as science shall endure. It is nothing less than the telescopic confirmation of the existence of this new planet, which had previously been indicated only by mathematical calculation.

Great indeed was the admiration of the scientific world at this superb triumph. Here was a mighty planet, whose very existence was revealed by the indications afforded by refined mathematical calculation. At once the name of Leverrier, already known to those conversant with the more profound branches of astronomy, became everywhere celebrated.

When, in 1854, Arago's place had to be filled at

the head of the great Paris Observatory, it was universally felt that the discoverer of Neptune was the suitable man to assume the office. Leverrier died on Sunday, September 23, 1877, in his sixty-seventh year.

FREDERICK WILLIAM BESSEL was born at Minden in 1784, and early turned his attention to mathematical subjects. He devoted himself with ardor to astronomy, and in 1804 he undertook the reduction of the observations made on the comet of 1607. His results were communicated to Olbers, who warmly praised the young astronomer, and in 1806 recommended him to Schroeter as an assistant in the observatory of Lilienthal. In 1810 he was appointed director of the new observatory then being founded by the king at Königsberg. He was admirably fitted for this post, and is distinguished mainly for his discovery of the parallax of the star 61 Cygni, which he accomplished by methods of extreme ingenuity and delicacy.

Two kinds of telescopes are commonly used, the refractor and the reflector. The first telescopes made were all refractors, and the very first of them was made, as you know, by Galileo. The first reflector was made in later days by Sir Isaac Newton.

In a refractor the rays of light, from a star or any other bright body, reach first a large object-glass, through which they pass, and by which, as they pass, they are caused to converge, narrowing to a focus or point. From this point they widen slightly on their way to the eye-piece. After passing through the eye-piece, by which they are once more straightened into parallel rays, they arrive at the eye.

Speaking broadly, the eye receives—actually sees—nearly as much of the starlight as if its pupil were the full size of the object-glass of that telescope, and the power of such an instrument depends upon the size of the object-glass. A refractor has been made with an object-glass forty inches across, and this, for the observer, means looking at a star with an eye-pupil more than three feet in diameter. This refractor is now in the Yerkes Observatory at Williams Bay, Lake Geneva, Wisconsin.

There are several forms of reflectors. With one form, when it is pointed at a star, the rays of starlight fall direct through the telescope-tube upon a highly-polished mirror, so shaped that rays of light reflected thence are caused to fall converging upon a small plane-mirror in the center of the tube, from which they are thrown, parallel to the side of the tube, to the eye-piece, which converges them to the eye.

The mirror of a reflector can be made very much larger than the object-glass of a refractor. The famous Lord Rosse telescope, which has a tube sixty feet long, contains a mirror no less than six feet in diameter. This, to an observer, is tantamount to looking at the stars with an eye so huge that its pupil alone would be nearly equal in size to the six-foot wheel of a steam-engine. It will readily be perceived how much more starlight can be grasped by such a fishing-net than by the tiny pupil of your eye or mine.

A main difference between the two kinds of telescope thus resides in the fact that the star-rays—or any other kind of light-rays—when first captured, pass, in one case, *through* the glass on which they fall, and, in the other case, are thrown off *from* the mirror. In both cases, the whole amount of light so

EYE-PIECE OF THE LICK TELESCOPE.

captured is gathered into a small compass, so as to be available for human sight.

Once more, let me remind you, it should be always kept clearly in mind that every object that is seen by us—from a mote of dust to a sun, from a coal-scuttle to a star—is perceived purely and solely by the light which it gives out, either intrinsic or reflected light. Countless myriads of rays pass from the surface of the thing seen to our eyes, picturing there, on the sensitive retina, a fleeting vision of its form.

All the leading Governments in Europe have observatories for the study of the heavens, which are furnished with the best instruments of modern construction. The principal observatories are those at Paris, Berlin, Vienna, Nice; Dorpat, the seat of a celebrated university founded by Gustavus Adolphus, of Sweden, in 1630; Pulkowa, near St. Petersburg; Lisbon, and at Greenwich in England. In America, though only in recent years has astronomy been cultivated with ardor, there are observatories of more or less importance,—the National Observatory at Washington, D. C.; the Cincinnati Observatory, projected by the late General O. M. Mitchel; the Dudley Observatory, founded by Mrs. Blandina Dudley in honor of her husband, at Albany, N. Y.; the Lick Observatory, founded by James Lick, on Mount Hamilton, Cal.; and the Yerkes Observatory, connected with the Chicago University, founded by Charles T. Yerkes. Besides these, there are observatories connected with some of our other leading universities and colleges, as those of Harvard, at Cambridge, Mass.; Yale University; Williams College; West Point Military Academy; Amherst College; Princeton; Dartmouth; Michigan University; Hamilton College, N. Y., and several others.

The largest and best refracting telescopes in the world are those at Lick Observatory and at Yerkes

LICK OBSERVATORY IN WINTER, FROM THE EAST.

Observatory, now in charge of Professor E. E. Barnard, who, while at the head of the Lick Observatory, discovered the fifth satellite of Jupiter, and determined the character of the inner ring of Saturn. The best

E. E. BARNARD.

reflecting telescope is that of Lord Rosse, at Birr Castle, in Kings County, Ireland.

In reviewing the path which we have traversed in this volume, though we have penetrated the sidereal depths and wandered at will through space, we feel that we have not yet reached the outskirts of creation. We have never left the center. Beyond us still lies the circumference. We see opened before us the infinite, of which the study is not yet begun! We have seen nothing; we recoil in terror; we fall back astounded. Indeed, we might fall into the yawning abyss—fall forever, during a whole eternity; never, never should we reach the bottom, any more than we have attained the summit. There is no east nor west, neither right nor left. There is no up nor down; there is neither a zenith nor a nadir. In whatever way we look, it is infinite in all directions.

In this infinitude of space, the associations of suns and of worlds which constitute our visible universe form but an island, a vast archipelago; and in the eternity of duration, the life of our proud humanity, with all its concerns, its aspirations, and its achievements, the whole life of our entire planet, is but the shadow of a dream! Well might the Hebrew poet, as he looked forth upon nature, exclaim in his amazement: "When I consider thy heavens, the work of thy fingers, the moon and the stars which thou hast ordained, what is man that thou art mindful of him, and the son of man that thou visitest him?"

The constellations, the charts of the sky; the catalogues of curious stars—variable, double or multiple, and colored; the description of instruments accessible to the observer of the heavens, and useful tables to consult, are only touched upon in this volume. The reader whose scientific desires are satisfied by the elements of astronomy, may stop here. Few have time to pursue the subject further; but it is sweet to live in the sphere of the mind; it is sweet to contemn the rough noises of a vulgar world; it is sweet to soar in the ethereal heights, and to devote the best moments of our life to the study of the true, the infinite, and the eternal!

PUBLISHERS' NOTE.

WE have now reached the conclusion of "The Story of the Sun, Moon, and Stars." Miss Giberne's work has been supplemented with a large amount of matter from other sources. The value of the book has thus been increased without diminishing its popular character. In most admirable form it tells the story of the heavenly world, which is the work of God's hands, the moon and the stars which he has made. The reverent reader, to whom the Word of God is "sweeter than honey and the honey-comb," may here find that "the heavens declare the glory of God, and the firmament showeth his handiwork." We confidently present it to the reader as the best and completest work in print on the great subject of which it treats.

INDEX.